ANIM
AT THE
CROSSROADS

PREVENTING, DETECTING, AND DIAGNOSING ANIMAL DISEASES

Committee on Assessing the Nation's Framework for Addressing Animal Diseases

Board on Agriculture and Natural Resources

Division on Earth and Life Studies

NATIONAL RESEARCH COUNCIL
OF THE NATIONAL ACADEMIES

THE NATIONAL ACADEMIES PRESS
Washington, D.C.
www.nap.edu

THE NATIONAL ACADEMIES PRESS • 500 Fifth Street, N.W. • Washington, D.C. 20001

NOTICE: The project that is the subject of this summary was approved by the Governing Board of the National Research Council, whose members are drawn from the councils of the National Academy of Sciences, the National Academy of Engineering, and the Institute of Medicine. The members of the committee responsible for the summary were chosen for their special competences and with regard for appropriate balance.

This report was supported by the National Academy of Sciences. Any opinions, findings, conclusions, or recommendations expressed in this publication are those of the author(s) and do not necessarily reflect the views of the organizations or agencies that provided support for the project.

International Standard Book Number 0-309-09259-0 (Book)
International Standard Book Number 0-309-533384 (PDF)
Library of Congress Control Number 2005932662

Additional copies of this report are available from the National Academies Press, 500 Fifth Street, N.W., Lockbox 285, Washington, D.C. 20055; (800) 624-6242 or (202) 334-3313 (in the Washington metropolitan area); Internet, http://www.nap.edu

Printed in the United States of America

THE NATIONAL ACADEMIES

Advisers to the Nation on Science, Engineering, and Medicine

The **National Academy of Sciences** is a private, nonprofit, self-perpetuating society of distinguished scholars engaged in scientific and engineering research, dedicated to the furtherance of science and technology and to their use for the general welfare. Upon the authority of the charter granted to it by the Congress in 1863, the Academy has a mandate that requires it to advise the federal government on scientific and technical matters. Dr. Ralph J. Cicerone is president of the National Academy of Sciences.

The **National Academy of Engineering** was established in 1964, under the charter of the National Academy of Sciences, as a parallel organization of outstanding engineers. It is autonomous in its administration and in the selection of its members, sharing with the National Academy of Sciences the responsibility for advising the federal government. The National Academy of Engineering also sponsors engineering programs aimed at meeting national needs, encourages education and research, and recognizes the superior achievements of engineers. Dr. Wm. A. Wulf is president of the National Academy of Engineering.

The **Institute of Medicine** was established in 1970 by the National Academy of Sciences to secure the services of eminent members of appropriate professions in the examination of policy matters pertaining to the health of the public. The Institute acts under the responsibility given to the National Academy of Sciences by its congressional charter to be an adviser to the federal government and, upon its own initiative, to identify issues of medical care, research, and education. Dr. Harvey V. Fineberg is president of the Institute of Medicine.

The **National Research Council** was organized by the National Academy of Sciences in 1916 to associate the broad community of science and technology with the Academy's purposes of furthering knowledge and advising the federal government. Functioning in accordance with general policies determined by the Academy, the Council has become the principal operating agency of both the National Academy of Sciences and the National Academy of Engineering in providing services to the government, the public, and the scientific and engineering communities. The Council is administered jointly by both Academies and the Institute of Medicine. Dr. Ralph J. Cicerone and Dr. Wm. A. Wulf are chair and vice chair, respectively, of the National Research Council.

www.national-academies.org

COMMITTEE ON ASSESSING THE NATION'S FRAMEWORK FOR ADDRESSING ANIMAL DISEASES

LONNIE J. KING, *Chair*, Michigan State University, East Lansing
MARGARET A. HAMBURG, *Vice Chair*, Nuclear Threat Initiative, Washington, D.C.
SHARON ANDERSON (Emeritus), North Dakota State University, Fargo
ALFONZA ATKINSON (deceased), Tuskegee University, Alabama
CORRIE BROWN, University of Georgia, Athens
TIMOTHY J. HERRMAN, Texas A&M University, College Station
SHARON K. HIETALA, University of California, Davis
HELEN H. JENSEN, Iowa State University, Ames
CAROL A. KEISER, C-BAR Cattle Company, Inc., Champaign, Illinois
SCOTT R. LILLIBRIDGE, The University of Texas Health Science Center, Houston
TERRY F. MCELWAIN, Washington State University, Pullman
N. OLE NIELSEN (Emeritus), University of Guelph, Spruce Grove, Alberta, Canada
ROBERT A. NORTON, Auburn University, Alabama
MICHAEL T. OSTERHOLM, University of Minnesota, Minneapolis
M. PATRICIA QUINLISK, Iowa Department of Public Health, Des Moines
LINDA J. SAIF, The Ohio State University, Wooster
MARK C. THURMOND, University of California, Davis
KEVIN D. WALKER, Inter-American Institute for Cooperation in Agriculture, Coronado, Costa Rica

National Research Council Staff

ROBIN SCHOEN, Study Director (since December 2004)
ELISABETH A. REESE, Study Director (July 2004 to December 2004)
TINA I. ROUSE, Study Director (through June 2004)
PEGGY TSAI, Research Associate (since November 2004)
TANJA PILZAK, Research Assistant (through July 2004)
DONNA WILKINSON, Research Intern

Acknowledgments

This report represents the integrated efforts of many individuals. The committee thanks all those who shared their insight and knowledge to bring the document to fruition. We also thank all those who provided information at our public meetings and who participated in our public sessions.

During the course of its deliberations, the committee sought assistance from many people who gave generously of their time to provide advice and information that the committee considered in its deliberations. Special thanks are due the following:

Bruce Akey, New York State Department of Agriculture and Markets
David Asher, United States Food and Drug Administration
John C. Bailar III, University of Chicago
Norman Crouch, Association of State Public Health Laboratories
Andrew Cupit, Embassy of Australia
Ron DeHaven, United States Department of Agriculture
Richard Dierks, Association of American Veterinary Medical Colleges
Leland Ellis, United States Department of Homeland Security
Brian Evans, Canadian Food Inspection Agency
Nathan Flesness, International Species Information System
Glen Garris, United States Department of Agriculture
Lawrence Heider, Association of American Veterinary Medical Colleges

Peter J. Johnson, United States Department of Agriculture
Elizabeth Krushinskie, Pilgrim's Pride Corporation
Karen E. Lawson, United States Department of Agriculture
Andrew Maccabe, Association of American Veterinary Medical Colleges
Curt Mann, White House Homeland Security Council
Maureen McCarthy, United States Department of Homeland Security
Thomas McKenna, United States Department of Agriculture
Lawrence E. Miller, United States Department of Agriculture/Veterinary Services
Andrea Morgan, United States Department of Agriculture/Veterinary Services
Mo Salman, Colorado State University
Scott Severin, Department of Defense Veterinary Service Activity
Nga Tran, Exponent, Inc. Food & Chemicals Practice
Leon Weaver, Bridgewater Dairy LLC
Gary Weber, National Cattlemen's Beef Association
Elizabeth Williams, University of Wyoming
Terry Wilson, United States Department of Agriculture

The committee is also grateful to members of the National Research Council staff who worked diligently to maintain progress and quality in its work, and to Paula Whitacre, for editing the report.

This report has been reviewed in draft form by individuals chosen for their diverse perspectives and technical expertise, in accordance with procedures approved by the National Research Council's Report Review Committee. The purpose of this independent review is to provide candid and critical comments that will assist the institution in making its published report as sound as possible and to ensure that the report meets institutional standards for objectivity, evidence, and responsiveness to the study charge. The review comments and draft manuscript remain confidential to protect the integrity of the deliberative process. We wish to thank the following individuals for their review of this report:

Alex Ardans, University of California, California Animal Health and Food Safety Lab
Nancy L. Ascher, University of California, San Francisco
Peter Eyre, Virginia Polytechnic Institute and State University
E. Paul J. Gibbs, University of Florida
George M. Gray, Harvard School of Public Health
Donald A. Henderson, Johns Hopkins University School of Hygiene and Public Health

Bob Hillman, Texas Animal Health Commission
Dennis F. Kohn, Columbia University College of Physicians and Surgeons
Gary Jay Kushner, Hogan & Hartson L.L.P
F.A. (Ted) Leighton, University of Saskatchewan
James D. McKean, Iowa Pork Industry Center
Harley Moon, Iowa State University
Suzanne Kennedy Stoskopf, Pylon Research Laboratories

Although the reviewers listed above provided many constructive comments and suggestions, they were not asked to endorse the conclusions or recommendations nor did they see the final draft of the report before its release. The review of this report was overseen by Linda Cork, Stanford University, and Mary Jane Osborn, University of Connecticut Health Center. Appointed by the National Research Council, they were responsible for making certain that an independent examination of this report was carried out in accordance with institutional procedures and that all review comments were carefully considered. Responsibility for the final content of this report rests entirely with the authoring committee and the institution.

This report is dedicated to the memory of Alfonza Atkinson, a member of the Committee on Assessing the Nation's Framework for Addressing Animal Diseases.

Preface

The committee was charged to assess the country's framework to support animal health in the context of our rapidly changing world and the contemporary challenges faced by those involved in animal health and diseases. In this report, the first of an intended three-part series, the committee members were asked to focus on the prevention, detection, and diagnosis of animal diseases and the dynamics of these systems as part of the overall animal health framework. It has also set the groundwork for two other studies and subsequent reports that will follow to assess the surveillance and response systems within the framework.

The world of animals—domestic, wildlife, and food-producing—and their health has increased in complexity and importance over the last century. In addition, the challenges and opportunities for animal health that have become especially apparent over the last several decades are unprecedented. Our animal health system is inextricably interwoven into both our national and global economy, as well as numerous societal issues including the public's health. Animal agriculture, in particular, finds itself in the midst of fundamental change and transforming forces. The scope, scale, and potential implications of the global food-animal system and its associated infrastructure to monitor and support animal health and food safety work is without precedent. In their examination of the animal health framework, with special reference to prevention, detection, and diagnosis, the committee members were struck with the new interdependence of animal health concerns and needs with issues such as public health and medicine, economics, global trade, and national and international security. Virtually all aspects of the current animal health framework are impacted by a related set of new challenges, relationships, and interactions that have emerged outside of traditional agricultural communities.

Because of its striking interdependence and connectiveness, animal health finds itself truly at a crossroads. This fact was made further evident as the committee examined contemporary threats and challenges, reviewed the framework from the retrospective analyses of recent disease events, and studied the gap between the current framework systems and what is needed for success. While the committee considered companion animals and wildlife as part of the animal health framework, this report emphasizes food-animals based on the nature and enormous challenges unique to this sector and the urgent need to address them.

The committee purposely had a significant human health component, which added greatly to its understanding and appreciation of the convergence of human and animal health and the strong linkage between animal and public health. The contemporary issues of emerging infectious diseases, new zoonoses, bio- and agroterrorism, antimicrobial resistant pathogens, and global health threats reaffirm the importance of the convergence and the consideration of these influences on the future of animal health and its associated framework.

The committee examined other reports and publications, listened to invited speakers, engaged in lengthy discussions, and brought together diverse perspectives and a variety of experts. Through this process and deliberations, a strong consensus developed from the current crossroads that the United States must pursue a very different path; the future of animal health and the prevention, detection, and diagnosis systems will have to be very different from the past. The animal health framework in the United States is ripe for a transformation characterized by improvements in capacity and skills, new strategic partnerships, integration of its work processes and systems, the understanding and adoption of new technologies, and a broader global perspective.

The U.S. animal population and its associated animal health framework represent an exceptional national asset that impacts the lives of people everyday. Yet, the very people whose lives are improved and who benefit from their relationship with animals and their products are progressively less aware, fail to perceive the relevancy, and consequently are seemingly less supportive of the animal health enterprise. This fact will add a significant burden to the needed transformation effort.

The National Academies convened this committee to assess and address the national animal health framework at a very special time in the history of animal health. The committee's findings and recommendations support the compelling need for significant changes to create a new future. The decisions made today will define this future, not decisions made tomorrow. The title of the report uses the analogy of a crossroads. This analogy suggests that multiple options and pathways exist to the future; however, the committee and its recommendations support the notion that an entirely new pathway needs to be created that will significantly change both the planners and implementers of the framework, and, most importantly, the ultimate destination.

Lonnie King, *Chair*
Margaret Hamburg, *Vice Chair*
Committee on Assessing the Nation's Framework for Addressing Animal Diseases

Contents

Tables, Figures, and Boxes

Tables

Figures

Boxes

Summary

SYNOPSIS

The national framework to safeguard animal health is of paramount importance to the U.S. economy, public health, and food supply. To strengthen the existing framework, the nation should establish a high-level, authoritative mechanism to coordinate interactions between the private sector and local, state, and federal agencies. New tools for detection, diagnosis, and risk analysis need to be developed now, and the capacity of the existing animal health laboratory network should be expanded for both routine and emergency diagnostic uses. Integrative animal health research programs, in which veterinary and medical scientists can work as collaborators, should be established. Colleges of veterinary medicine must lead an effort to develop a national animal health education plan to educate and train individuals from all sectors (from animal handlers to pathologists) in disease prevention and early detection and to recruit veterinary students into careers in public health, food systems, biomedical research, diagnostic laboratory investigation, pathology, epidemiology, ecosystem health, and food-animal practice. The United States must address the importation and health of exotic and wild-caught animals and commit itself to shared leadership roles with other countries and international organizations that address animal disease agents. Finally, a collective effort should be made to raise the level of public awareness about the importance of animal health and of the national investment in the framework to safeguard animal health.

BACKGROUND

Animal health has broad implications, ranging from the health of individual animals and the well-being of human communities to issues of global security. Many people would be surprised by the assertion that our nation's highest priorities must include animal health, yet we must recognize and act on this reality to ensure a safe and healthy future. Among other things, animal diseases critically affect the adequacy of the food supply for a growing world population, and they have huge implications for global trade and commerce. Moreover, many animal disease agents are zoonotic—meaning that they are transmittable to humans—so they have dramatic implications for human health and safety, and for animal disease prevention. Animal disease prevention and control is crucial to improving public health on a global scale. In addition, in an era of growing concern about the threat of terrorism, the potential impact of the intentional use of animal disease agents to cause morbidity and mortality, as well as economic damage, is enormous.

The U.S. animal health framework includes many federal, state, and local agencies that generally have differing mandates of law and numerous other public and private entities and international organizations, each with its own goals and objectives, each responsible for maintaining animal health. In the past, this framework has been reasonably effective in responding to a range of demands and challenges. In recent years, however, animal health has been challenged in a manner not previously experienced.

Today animal health is at a crossroads. The risk of disease is coming from many directions, including the globalization of commerce, the restructuring and consolidation of global food and agriculture productions into larger commercial units, the interactions of humans and companion animals, human incursions into wildlife habitats, and the threat of bioterrorism. The impacts of these sources of risk are evident in recent disease events (Box S-1).

Given the changing nature of the risks with which the framework must cope, it is unlikely that the current philosophy on how to protect animal health will be adequate in the future. The risks of animal disease must be dealt with not only in terms of protecting individual species of animals from specific pathogens, but also in a broader context that includes anticipating the emergence and spread of disease on local and global scales and recognizing the relationships of animal disease to human health and the environment. To address animal disease in that context, the animal health framework will have to be more flexible and inclusive of expertise available from research, medical, and public health communities, and from the fields of environmental sciences and public policy, among others. To respond comprehensively to new threats, the responsi-

BOX S-1
Impacts from Recent Disease Events

- In 2003, severe acute respiratory syndrome (SARS) sent a global shock wave, affecting countries with even few cases, such as the United States. Although SARS infected only 8,000 people globally, the disease spread to 30 countries and its effect on the global economy totaled $8 billion.

- The United Kingdom's economy has not yet recovered from a foot-and-mouth disease (FMD) outbreak in 2001, which also reverberated around the world, affecting both agricultural and nonagricultural interests (such as rural businesses and tourism/recreational use of the countryside).

- A single case of mad cow disease (bovine spongiform encephalopathy or BSE) in Washington State on December 23, 2003, had an immediate market impact and severe, sustained economic losses due to trade restrictions on U.S. cattle and their products. The infected animal was discovered as part of the government's policy to routinely test downer cattle for BSE, which has been linked to a new variant of Creutzfeldt-Jakob disease, a fatal neurological illness in humans. In June 2005, a second case of BSE was confirmed in the United States.

- In 2004, a new strain of highly pathogenic avian influenza (AI) spread through Southeast Asia, resulting in loss of more than 100 million birds through mortality and control measures and dozens of human cases, highlighting the unpredictable and potentially catastrophic nature of an emerging zoonotic disease. This new influenza strain was transmitted from birds to people, raising concern that it might be capable of evolving into the next pandemic influenza strain.

- In 1999, West Nile virus (WNV), an arbovirus similar to St. Louis encephalitis virus, emerged for the first time in the Western Hemisphere in New York from an unknown source. Over the next five years it swept across the continental United States, Canada, Mexico, Central America, and several Caribbean islands, carried by mosquito vectors infecting wild birds. In the United States in 2004, the virus was detected in approximately 2,250 humans (40 states), 1,250 horses (36 states), nearly 7,000 wild birds, mostly corvids (45 states), and in much smaller numbers in a few other animal species. While these numbers are substantially below those that occurred in the first wave of infection, WNV bodes to become endemic in wild birds and an ongoing source of infection transmitted to other species by mosquito vectors.

bilities of the framework's many actors will need to be clearly defined and their actions better coordinated. Admittedly, the process of transformation is difficult during periods when disease outbreaks consume all attention. However, *now* is the time to strengthen the structure of the current system and to instigate a change in its culture, so that it will be capable of responding effectively in the future.

This report explores the evolving challenges facing animal health, identifies vulnerabilities and gaps in the animal health framework, and recommends steps needed to fill gaps and improve the effectiveness of the framework.

COMMITTEE'S STATEMENT OF TASK

Recent animal and human health events have illustrated that the national system for protecting animal health is now facing a continuum of host-parasite relationships involving public health, wildlife, ecosystems, and food systems, operating in an increasingly complex global context (see Figure S-1). Adapting the current framework to this new reality will be both a major challenge and a national imperative.

In recognition of the changing influences on animal health, the National Academies developed a concept for a three-phase analysis of the U.S. system for dealing with animal diseases and committed institutional funds to launch the first phase of the study. This report, which embodies the first phase of the study, presents an overview of the animal health framework and examines the framework's overall operation in the prevention, detection, and diagnosis of animal diseases. The proposed second phase of the study (pending supplemental external support) will focus on surveillance and monitoring capabilities, and the proposed third phase will focus on response and recovery from an animal disease epidemic. Although surveillance and monitoring play an important part in prevention, detection, and diagnosis, the second phase of the study, as currently envisioned, will analyze in greater depth the system's capacity and needs for surveillance and monitoring of animal diseases.

Relative to its respective focus, each phase of the study will: (1) review the state and quality of the current system for dealing with animal disease; (2) identify key opportunities and barriers to successfully preventing and controlling animal diseases; and (3) identify immediate courses of action for those on the front lines.

This first phase of the study did not attempt an in-depth review of the effectiveness of each individual component of the framework or of any specific agency involved in safeguarding animal health—a task well beyond the scope of this effort—but did examine the effectiveness of the

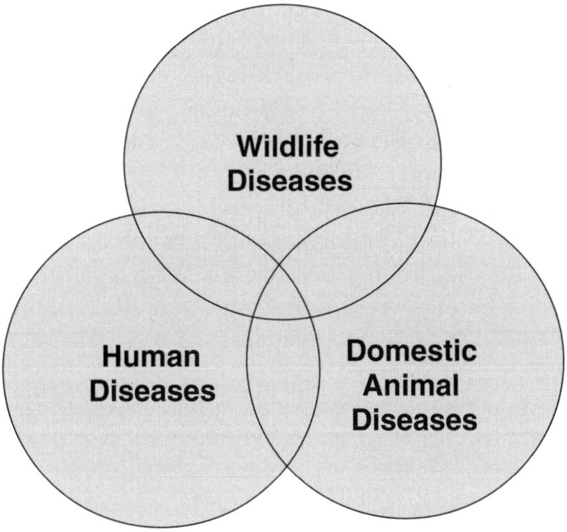

FIGURE S-1 Interactions of emerging infectious diseases (EIDs) with a continuum that includes wildlife, domestic animal, and human populations. Few diseases affect exclusively one group, and the complex relations among host populations set the scene for disease emergence. Examples of EIDs that overlap these categories include Lyme disease (wildlife to domestic animals and humans); bovine tuberculosis (between domestic animals and wildlife); *Escherichia coli* O157:H7 (between domestic animals and humans); and Nipah virus and rabies (all three categories). Companion animals are categorized in the domestic animal section of the continuum.

framework as a whole in relation to different animal disease scenarios. In doing so, it sought to identify ways to improve the framework.

Finally, although animals are subject to the same causes of disease as humans—that is, diseases with chemical, physical, microbial, or genetic causes—the study focuses primarily on infectious diseases, as directed by the Statement of Task (see Chapter 1, Box 1-1, for the committee's Statement of Task). This focus arises from concern about the growing threat posed by the spread of emerging infectious disease associated with the increasing global interconnectedness of domestic animals, wildlife, and humans, and by the possibility of bioterrorism.

OVERVIEW OF THE ANIMAL HEALTH FRAMEWORK

The essential components of the animal health framework include the following:

- people on the "front lines" of the animal production unit, animal habitat, or companion animal household (including ranch and farm workers, producers, feeders, breeders, park rangers, companion animal owners, wildlife rehabilitators, and zoo keepers);
- veterinarians and other sources of professional advice and care for health-related issues (such as universities and diagnostic laboratories);
- federal, state, and local animal health and public health agencies (consisting mainly of state departments of agriculture and state diagnostics laboratories within universities and elsewhere in state governments, and numerous bureaus and offices within over 10 federal departments, but primarily within the U.S. Departments of Agriculture, Homeland Security, and Health and Human Services);
- international collaborations among agencies, organizations, and governments (such as the World Organization for Animal Health and the World Health Organization);
- supporting institutions, industries, and organizations (including educators, researchers, and the public health and intelligence communities).

Because of the very large number of actors responsible in some way for safeguarding animal health, it is not surprising that effective coordination is a major challenge. In a retrospective analysis of numerous specific animal disease situations, the committee examined the collective capabilities and limitations of the framework with respect to its effectiveness in the prevention, detection, and diagnosis of animal diseases. Several weaknesses, needs, and gaps were consistently encountered in the framework's response to a broad spectrum of disease types including exotic Newcastle disease (END), foot-and-mouth disease (FMD), monkeypox, bovine spongiform encephalopathy (BSE), chronic wasting disease (CWD), West Nile virus (WNV), avian influenza (AI), and diseases caused by coronavirus. This examination led the committee to the following conclusions:

- The framework for animal health lacks adequate systems and tools for analyzing and managing risk, and planning for outbreaks.
- Efforts to develop and validate diagnostic assays and advanced vaccines of a recognized pathogen need to occur more rapidly.
- The workforce on the front lines of animal care is not adequately educated and trained to deal with animal disease issues, and there is a shortage of veterinarians in the workforce for animal disease prevention, detection, and diagnosis.
- Greater collaboration between public health and animal health officials can accelerate the detection and diagnosis of animal diseases.

- The broad capabilities that exist in universities, industry, state entities, veterinary diagnostic laboratories, and other local animal health infrastructure are underutilized.
- The lack of collaboration between the biomedical and veterinary communities is a lost opportunity that impedes the effectiveness of the framework.
- There is a need for state-of-the-art equipment and biocontainment facilities for both research and diagnostics. Federal, state, and private entities responsible for animal health have different authorities, and there are gaps in that authority, particularly in relation to wildlife disease.
- The past success of international collaboration in responding to animal disease demonstrates its importance in addressing animal diseases.

RECOMMENDATIONS FOR STRENGTHENING THE ANIMAL HEALTH FRAMEWORK

Reflecting on the structure of the framework and based on the findings of its analysis of past animal health events, the committee offers the following 11 recommendations as potential opportunities for strengthening the framework's capabilities in the prevention, detection, and diagnosis of animal diseases.

Coordination of Framework Components

Recommendation 1: **The nation should establish a high-level, centralized, authoritative, and accountable coordinating mechanism or focal point for engaging and enhancing partnerships among local, state, and federal agencies and the private sector.**

There is a need for a strategic focal point to enhance partnerships and to integrate all stakeholders into a cohesive whole. Many federal agencies are responsible for parts of animal health policy, with significant overlaps in the programmatic functions among them and also between federal agencies and programs directed through states or animal health organizations. On the other hand, there are also gaps in responsibility. Of particular concern is the paucity of federal oversight of the nonlivestock, animal-centered aspects of zoonotic diseases.

While there are several possible models for improved coordination in prevention, detection, and diagnosis, the committee did not recommend options for a specific system-wide mechanism, in part because it has only examined the animal health framework from the partial perspective of prevention, detection, and diagnosis.

Regardless of how a central coordinating mechanism or focal point is implemented, it will need to promote effective communication among various stakeholders and with the public during and outside episodes of animal disease outbreaks. Opportunities for information-sharing between agencies using electronic information systems should be developed. A methodic effort should be made to identify and link key databases and establish protocols for contributing data and generating alerts.

Technological Tools for Preventing, Detecting, and Diagnosing Animal Diseases

Recommendation 2: **Agencies and institutions—including the U.S. Department of Agriculture (USDA) and the Department of Homeland Security (DHS)—responsible for protecting animal industries, wildlife, and associated economies should encourage and support rapid development, validation, and adoption of new technologies and scientific tools for the detection, diagnosis, and prevention of animal diseases and zoonoses.**

The current animal health framework has been slow to evaluate, validate, and implement new scientific tools and technologies that could significantly enhance animal disease prevention, detection, and diagnostic capabilities for the United States. Despite a recent surge in activity related to post-September 11 homeland security efforts and associated focused funding, the active review and implementation of advancing technologies has been inadequate to protect and enhance the health of the country's animal populations and related economic systems. Technological advances—such as immune system modulators, animal-embedded monitoring (chips embedded underneath an animal's skin to monitor temperature and other physiological indices), and differential vaccines as prevention strategies, as well as a range of rapid, automated, sensitive, and portable sampling and assay systems for early warning and reliable diagnosis—have not been adequately exploited by the current animal health framework. Early biodefense warning systems, such as DHS' BioWatch or private industry's gene-based anthrax testing, are designed for rapid detection and identification of key pathogens by sampling air in public environments. These systems have been operating since early 2003 and are meant to assist public health experts in rapidly responding to the intentional release of a biologic agent (DHS, 2004a). Early warning technologies have not yet been adequately evaluated by the animal health infrastructure.

Scientific Preparedness for Diagnosing Animal Diseases: Laboratory Capacity and Capability

Recommendation 3: **The animal health laboratory network should be expanded and strengthened to ensure sufficient capability and capacity for both routine and emergency diagnostic needs and to ensure a robust linkage of all components (federal, state, university, and commercial laboratories) involved in the diagnosis of animal and zoonotic diseases.**

Laboratory diagnosis of animal diseases in the United States involves federal, state, university, and commercial entities. The committee focused its assessment on the condition of publicly funded laboratories and the current operational status of national laboratory networks. Funding and implementation of the pilot National Animal Health Laboratory Network (NAHLN) in June 2002 was an important and beneficial paradigm shift from an exclusive federal system to one with shared state and federal responsibility for foreign animal disease diagnosis. The pilot NAHLN involved 12 state/university diagnostic laboratories approved for disease testing using existing and newly developed assays. The NAHLN is no longer a pilot program and has since been redefined to include all laboratories performing contract work for the USDA on BSE, CWD, scrapie, AI, END, and classical swine fever (CSF). However, the current network lacks surge capacity and is not prepared for disease agents and toxins outside the narrow list of diseases that provided an initial focus for network development (for example, FMD or Rift Valley fever). In addition, implementing this recommendation will require the creation of formal linkages and operational relationships between the NAHLN, state and university veterinary diagnostic laboratories, and the Laboratory Response Network for Bioterrorism (LRN), established by the U.S. Centers for Disease Control and Prevention (CDC) in 1995 to improve the response capabilities of the nation's public health laboratory infrastructure. It will require development of additional biosafety level 3 (BSL-3) necropsy and laboratory capacity. Population-based diagnostic and detection systems also will need to be developed by diagnostic laboratories in order to provide the broad diagnostic outlook necessary for detection of new and emerging diseases.

Animal Health Research

Recommendation 4: **Federal agencies involved in biomedical research (both human and veterinary) should establish a method to jointly fund new, competitive, comprehensive, and integrated animal health research programs; ensure that veterinary and medical scientists can work as collaborators; and enhance research, both domestically and interna-**

tionally, on the prevention, detection, and diagnosis of animal and zoonotic disease encompassing both animal and human hosts.

This process might be modeled on the National Institutes of Health (NIH)-administered Interagency Comparative Medicine Research Program, an interagency task force model, or some comparable process that promotes this type of cooperative research agenda.

This recommendation builds on the 2003 Institute of Medicine (IOM) report *Microbial Threats to Health: Emergence, Detection, and Response,* which states: "NIH should develop a comprehensive research agenda for infectious disease prevention and control in collaboration with other federal research institutions and laboratories (e.g., CDC, the U.S. Department of Defense, Department of Energy, the National Science Foundation), academia, and industry" (IOM, 2003). Currently, basic and translational research related to prevention, detection, and diagnosis of animal and zoonotic diseases is being conducted by a complex array of government, academic, and private institutions and there is minimal coordination, if any, in setting priorities to ensure that important research topics are not overlooked and to ensure the most effective use of scarce resources. A forthcoming National Research Council (NRC) report *Critical Needs for Research in Veterinary Science* will contain a more in-depth assessment of national needs for research in animal health.

Recommendation 5: **To strengthen the animal health and zoonotic disease research infrastructure, the committee recommends that competitive grants be made available to scientists to upgrade equipment for animal disease research and that the nation construct and maintain government and university biosafety level 3 (BSL-3 and BSL-3 Ag)[1] facilities for livestock (including large animals), poultry, and wildlife.**

Access to state-of-the-art equipment and technological tools is essential to successfully conduct the research needed to understand, prevent, and control emerging or exotic infectious agents. When a new infectious agent is suspected, efforts must be made to first rapidly define and characterize the agent, under strict biocontainment conditions. At present, few BSL-3 or BSL-3 Ag facilities are available strategically throughout the United States or are equipped and prepared for research on diseases of livestock, poultry, or wildlife, including zoonoses that require BSL-3

[1]Containment facilities are classified as Biosafety Levels 1 through 4, with 4 being the most restrictive. Biosafety level 3 (BSL-3 or BSL-3 Ag) provides the high degree of containment that is needed when studying a variety of organisms with a recognized potential for significant detrimental impact on animal or human health or on natural ecosystems.

biocontainment. Additional BSL-3 facilities are needed for research and surge capacity (in case of outbreaks) for detection, diagnosis, and prevention of many zoonotic and all exotic animal pathogens.

International Interdependence and Collaboration

Recommendation 6: **The United States should commit resources and develop new shared leadership roles with other countries and international organizations in creating global systems for preventing, detecting, and diagnosing known and emerging diseases, disease agents, and disease threats as they relate to animal and public health.**

As the United States and the rest of the world become increasingly interdependent, it is essential to identify animal disease risk factors as they emerge and to focus more attention on the sources and precursors of infections. Taken collectively, the recent experience with SARS, West Nile virus, and monkeypox leads to the inescapable conclusion that globalization, population growth, and expansion of human activity into previously unoccupied habitats has essentially connected the United States to potential zoonotic and nonzoonotic pathogens residing throughout the world. This necessitates coordinated international collaboration efforts directed at identifying potential risks worldwide, including regulatory mechanisms that minimize the threat of introducing emerging infectious agents into the United States or other unaffected countries.

For potential and emerging infectious agents in other countries, assistance from the United States is more ad hoc or piecemeal than strategic and wide-ranging. By adopting a more comprehensive approach to helping countries strengthen their prevention, detection, and diagnostic capabilities, the United States will enhance its own animal health framework and security. Means to accomplish this include transferring technology between nations and providing training opportunities to international students and veterinarians to ensure self-sufficiency and sustainable surveillance. The United States can also encourage and support the enhancement of critical competencies within the national services, which includes active participation in the formulation of international standards and the timely reporting of zoonotic and exotic diseases. The charge to the committee explicitly states that it will "review the U.S. system and approach for dealing with animal diseases," and the committee regards the international dimension as an extremely critical component of the U.S. animal health framework. With increased globalization and movement of diseases, people, products, pathogens, and vectors, the United States cannot continue to impose a line between domestic and international issues but should instead adopt an animal health system that identifies and responds to animal disease threats without regard to national boundaries.

Importation, Sale, and Transport of Animals

Recommendation 7: **Integrated and standardized regulations should be developed and implemented nationally to address the import, sale, movement, and health of exotic, nondomesticated, and wild-caught animals.**

Such a policy development needs to include health professionals and laboratory-based analysis because wild-caught and exotic animals may carry pathogens and pose a risk of transmitting disease without demonstrating clinical signs. The monkeypox outbreak of 2003 highlighted a number of weaknesses in the animal health framework for addressing a newly emergent zoonotic disease. In particular, while several federal agencies (including the U.S. Department of Agriculture, U.S. Department of the Interior's Bureau of Fish and Wildlife Service, and the U.S. Department of Health and Human Services) have roles in preventing, detecting, and diagnosing zoonotic and other diseases transmitted by exotic animals, there is a lack of coordinated federal oversight of the animal-centered aspects of diseases transmitted by exotic animals. Prior to the interim final rule banning the import, sale, or distribution of prairie dogs and some African rodents (responsible for the monkeypox outbreak in 2003), import and movement of exotic animals was largely uncontrolled (and most exotic animal movement is still uncontrolled). Tracking of these animals in the United States is inconsistent and ineffective, and there is a disturbing lack of standardized testing of the health status of exotic animals at the point of origin and in companion animal shops, trade fairs, and other venues. Considering that the emergence of new disease agents occurs most frequently at species interfaces, monkeypox is not likely to be the last zoonotic agent to emerge from an exotic animal in the United States.

Addressing Future Animal Disease Risks

Recommendation 8: **The USDA, DHS, Department of Health and Human Services, and state animal and public health agencies and laboratories should improve, expand, and formalize the use of predictive, risk-based tools and models to develop prevention, detection, diagnostic, and biosecurity systems and strategies for indigenous, exotic, and emerging animal diseases.**

There has been increased recognition and use of well-structured and scientifically based mathematical, epidemiological, and risk analysis models and tools to define acceptable risks and mitigation strategies that can assist in policy and science-based decision making. Examples include models of the spread of FMD during the U.K. epidemic, and an assessment of the risk of BSE to U.S. agriculture, developed by Harvard

University's Center for Risk Analysis for the USDA (Cohen et al., 2003; Haydon et al., 2004). Risk analysis and modeling have been criticized, mainly on the basis of insufficient scientific data or inappropriate assumptions. Therefore, efforts to develop scientific data on disease transmission, effectiveness of control programs, economic evaluation, and quantitative assessment of all factors involved in making policies and regulations should be a priority of the animal health infrastructure, working in collaboration with academia, industry, and global trade partners.

Threats from bioterrorism, emerging diseases, and foreign animal disease introductions add urgency to preventing or minimizing catastrophic consequences to the United States, other nations, and the global economy. Education and training of professionals to assess, manage, and communicate risk of animal disease and improved information available to stakeholders, including producers and the public, are important aspects of effective infrastructure that supports risk-based approaches.

Education and Training

Recommendation 9: **Industry, producers, the American Veterinary Medical Association (AVMA), government agencies, and colleges of veterinary medicine should build veterinary capacity through both recruitment and preparation of additional veterinary graduates into careers in public health, food systems, biomedical research, diagnostic laboratory investigation, pathology, epidemiology, ecosystem health, and food animal practice.**

There are insufficient graduates to meet the needs in a number of major and distinct fields of veterinary medicine dealing with various species of food-animals, rural practice (mixed domestic animals), ecosystem health (including wildlife disease and conservation biology), public health, the many dimensions of the food system, and biomedical science. In addition, veterinary graduates are not adequately prepared to deal with foreign animal diseases, public health, and ecosystem health, without further postgraduate studies. According to the Association of American Veterinary Medical Colleges (AAVMC), the 28 veterinary colleges in the United States graduate approximately 2,300 veterinarians per year and are currently unable to keep up with societal needs in private or public practice.

There has been a steady decline in the number of rural practitioners and of veterinarians employed in regulatory agencies. The USDA, underserved at present, predicts a shortfall of 584 veterinarians on its staff by 2007. Fifty percent of U.S. Public Health Service veterinarians are currently eligible for retirement.

Too few veterinary students are choosing to specialize in basic biomedical science or pathology, as noted in the recently published NRC report *National Need and Priorities for Veterinarians in Biomedical Research,* which suggests a strategy for recruiting and preparing more veterinarians for careers in laboratory animal medicine, comparative medicine, and comparative pathology (NRC, 2004b). This committee endorses the recommendations of that report.

One strategy for building veterinary capacity is to design and implement training and educational curricula to better address these underserved areas of animal health. The Veterinary Workforce Expansion Act of 2005, which amends the Public Health Service Act, will be a useful first step that establishes a competitive grants program to build capacity in veterinary medical education and expands the workforce of veterinarians engaged in public health practice and biomedical research.

Recommendation 10: **The USDA, state animal health agencies, the AVMA, and colleges and schools of veterinary medicine and departments of animal science should develop a national animal health education plan focusing on education and training of individuals from all sectors involved in disease prevention and early detection through day-to-day oversight of animals.**

Responsibility for implementing the educational plan would fall on those at the local level. Strong and well-functioning front-line detection is provided by animal handlers and personnel working with animals on a day-to-day basis. This backbone for effectively preventing animal disease outbreaks requires education and training to include awareness and recognition of clinical signs, as well as an elementary understanding of disease transmission and prevention. In addition, those with day-to-day oversight of animals need to understand the methods and responsibilities for reporting the signs of foreign and exotic animal diseases. Basic multilingual education and training are necessary for those with such direct oversight of animals, whereas more in-depth education to promote a greater depth and breadth of understanding of transmission and prevention is required for managers and owners.

Improving Public Awareness of the Economic, Social, and Human Health Effects of Animal Diseases

Recommendation 11: **The government, private sector, and professional and industry associations should collectively educate and raise the level of awareness of the general public about the importance of public and private investment to strengthen the animal health framework.**

Increased public awareness is critical in supporting and implementing transformations needed to strengthen the framework against animal disease risks. The lack of cohesive national advocacy for public health issues generally creates a much more difficult environment in which to increase attention and investment in the framework for preventing, detecting, and diagnosing animal diseases.

The recent outbreaks of FMD, SARS, AI, and BSE are all reminders of the threats such diseases pose to the U.S. food supply, global economy, public health, and confidence in the safety of the food supply. The entire food and fiber system—including farm inputs, processing, manufacturing, exporting, and related services—is one of the largest sectors of the U.S. economy and accounts for output of over $2 trillion dollars, generating $1.24 trillion in added value, and 12.3 percent of total gross domestic product in 2001 (USDA, 2003). Nearly 17 percent of all U.S. workers are employed by the food and fiber system (USDA, 2003). Producers, companion animal owners, and others on the front line have a direct personal and private interest in detecting, diagnosing, and preventing animal diseases to avoid losses associated with reduced productivity, animal mortality, or potential effects on personal health and welfare. Although these losses can be significant, adverse social, economic, and human health impacts associated with animal diseases extend beyond producers or household animal owners.

Increased investment in educating the public about animal health will help to reduce disease and transmission; enhance public and animal health; ensure a secure, economical, and viable food supply; and improve trade and competitiveness. These educational efforts should include information about diseases of food-animals, wildlife, and companion animals.

1

Introduction

"In the highly interconnected and readily traversed 'global village' of our time, one nation's problem soon becomes every nation's problem..."

—*Microbial Threats to Health: Emergence, Detection, and Response, Institute of Medicine, March 2003*

THE COMMITTEE'S STATEMENT OF TASK

Animal health is profoundly affected by the forces of globalization and trade, the threat of bioterrorism, the restructuring and consolidation of food and agriculture production into increasingly larger commercial units, and even by human incursions into wildlife habitats. A very large network of people, organizations, and operations undergird a framework of systems to protect animal health in the face of these forces; when changes occur that affect animal health, they also impact the framework. In recognition of the importance of the relationship of the animal health framework to changing conditions for animal health domestically and internationally, the National Academies and its Board on Agriculture and Natural Resources (BANR) have launched the first phase of a proposed, three-part initiative to review the state and quality of the framework and evaluate current opportunities and challenges to its effectiveness to preserve animal health (see Box 1-1).

The Committee's Approach to Its Task

As the product of the first stage of the initiative, this report provides a general overview of the structure of the animal health framework; identifies opportunities and barriers to the *prevention, detection,* and *diagnosis* of animal diseases; and, recommends courses of action for first-line responders and other participants in the framework, including the potential to apply new scientific knowledge and tools to address disease threats. This

BOX 1-1
Study Overview and Statement of Task for Phase One

A comprehensive review of the U.S. system and approach for dealing with animal diseases will be conducted. This initiative will (1) review, summarize, and evaluate the state and quality of the current system and the potential for improved application of scientific knowledge and tools to address threats and response efforts; (2) identify key opportunities and barriers to successfully preventing and controlling animal diseases as they relate to responsibilities and actions of producers, regulators, policy makers, and animal health care providers; and (3) identify courses of action for first-line responders to integrate strengths of proven strategies with promising approaches to meet animal health and management challenges. The study will be conducted in three phases, which correlate to the three major components of the U.S. structure of defense against animal diseases outlined in previous Board on Agriculture and Natural Resource (BANR) reports. The first phase of study will focus on the nation's framework for prevention, detection, and diagnosis of animal disease. The second phase will focus on the nation's system of monitoring and surveillance, and the third phase will focus on mechanisms of response and recovery from an animal disease epidemic. A core group of committee members will be appointed to participate in all phases of the activity to ensure consistency among the different phases, supplemented with additional expertise as needed for each phase.

In its examination, the committee will assess the adequacy of:

- Scientific preparedness for action "on the ground"
- Technologic tools and scientific applications
- Social and economic effects
- Reporting linkages and communications
- System components (federal, state, local, public and private)
- Supporting systems (research, education, and training).

For the first phase of the study, the committee will examine challenges in prevention, detection, and diagnosis presented by at least two specific animal diseases such as rinderpest; foot-and-mouth disease; West Nile virus; avian influenza; Newcastle disease; spongiform encephalopathies (scrapie in sheep and goats, chronic wasting disease [CWD] in deer and elk, transmissible mink encephalopathy [TME], and feline spongiform encephalopathy), and Q fever. These diseases represent a sample of diseases categories that have potential economic impact, human and/or animal health impact, or are foreign to the United States. They represent diseases, some which are zoonotic, that could affect each of the major agricultural species. The study will not address diseases that have been recently studied by BANR, such as brucellosis, Johne's disease, or bovine tuberculosis.

continued

BOX 1-1 Continued

For the diseases (or disease categories) it examines, the committee will assess the state of knowledge of each disease and its potential to cause animal health, human health, and social or economic impacts. The committee will review the etiology of the disease, the nature of the responsible pathogen(s), evidence and mechanisms of intra- and interspecific transmission of the diseases, and currently available and potential methods of diagnostic testing. Domestic and foreign approaches to prevention, detection, and diagnosis will be examined. For this initial phase of the review, recommendations will be provided on how to improve the nation's ability to address animal diseases by reducing potential for intentional or accidental introduction, enhancing diagnostic techniques and their use, and improving detection capabilities. Knowledge gaps and future needs for progress in systems and policies will be identified.

report lays the foundation for two additional phases proposed for the study of the animal health framework (surveillance and monitoring in phase 2 and response and recovery in phase 3).

Given the complex, global, and sometimes rapidly changing nature of events affecting the animal health framework, the committee looked beyond farming and food-producing animals to consider a broader array of topics and players. For example, because of the emergence of new zoonoses with transmission routes through wildlife, such as West Nile virus, the committee's approach includes threats to human and animal health from those sources. Beyond health concerns, the report also considers societal issues affected by animal disease outbreaks, such as economic impacts and food security. As a result, the report addresses a wide range and diversity of specific diseases from acute to chronic, endemic to exotic, and considers how naturally-occurring to intentionally spread might be handled. Box 1-2 presents a list of specific diseases examined in this report, which were selected to elaborate the need for an inclusive animal health infrastructure capable of preventing, detecting, and diagnosing a wide variety of animal health events.

Stakeholders with diverse perspectives involved in the animal health framework are the target audiences for this report, including animal producers, veterinarians, academic animal health educators and researchers, laboratory diagnosticians, state and federal elected officials, the public health community, state/local government officials, the technical community, policymakers, and the general public.

BOX 1-2
Animal Diseases Addressed in This Report

- Exotic (Foreign) Animal Diseases
  - Exotic Newcastle disease (END)
  - Foot-and-mouth disease (FMD)
- Recently Emergent Diseases
  - Monkeypox
  - Bovine spongiform encephalopathy (BSE)
- Endemic Diseases
  - Chronic wasting disease (CWD)
  - West Nile virus (WNV)
  - Avian influenza (AI)
- Previously Unknown Agents
  - Severe acute respiratory syndrome (SARS) coronavirus
- Novel Naturally Occurring or Bioengineered Animal Pathogens
- Diseases of Toxicological Origin

Boundaries of the Report

This report is not a comprehensive, all-inclusive discussion of animal health. Rather, it is intended as a first step to begin an analysis that will be expanded and enriched in the second and third planned phases of this study.

For example, this phase of the study did not attempt an in-depth review of the effectiveness of each individual component of the framework or of any specific agency involved in safeguarding animal health but did examine the effectiveness of the framework as a whole in relation to different animal disease scenarios and, in doing so, sought to identify ways that the framework could be improved.

Early in its deliberations, the committee found that the topics of the first phase—prevention, detection, and diagnosis—are intimately intertwined with surveillance, monitoring, and response/recovery issues and that they are impossible to deal with in isolation. Because surveillance and monitoring are important to the prevention, detection, and diagnosis of disease, they are referenced in the report. However, this volume has not reviewed and presented recommendations for an overall surveillance system or the infrastructure that might be required for such as system, because those issues are within the purview of the proposed second phase of this study.

General Terminology

This report contains numerous technical terms and acronyms or abbreviations (see Appendix A for Acronyms and Abbreviations; Appendix B for Glossary of Terms). Some of the general terms used frequently throughout the report are provided in Box 1-3, and a couple of them warrant additional clarification. The Statement of Task (Box 1-1) refers to "first-line responders"—individuals who play a key role in disease detection. Throughout this report, these individuals are identified as those on the front lines of detection, as described in Box 1-3.

BOX 1-3
General Terminology Used in This Report

Animal Health Framework: The collection of organizations and participants in the public and private sectors directly responsible for maintaining the health of all animals impacted by animal disease or that influence its determinants.

Front-Line Detection: Almost anyone can play a role in front-line detection and prevention (e.g., a school bus driver who notices a sick animal in a nearby field), but front-line detection and prevention as used in this report refers specifically to those in a position most likely to be the first judge of an abnormal health situation in an animal or population and to initiate preventive action. They include people involved directly in animal production as well as field personnel involved in wildlife management. Those with close and direct animal contact and observation include ranch and farm workers, feeders, breeders, milkers, animal sales yard personnel, slaughterhouse inspectors, dealers, park rangers, zoo keepers, and companion animal owners.

Exotic Animal Disease: Any animal disease caused by a disease agent that does not naturally occur in the United States (e.g., SARS, monkeypox).

Foreign Animal Disease: An exotic animal disease limited to agricultural animals (e.g., foot-and-mouth disease, bovine spongiform encephalopathy, rinderpest).

Zoonoses: Diseases caused by infectious agents that can be transmitted between (or are shared by) animals and humans.

The terms "exotic animal disease" and "foreign animal disease" are also used in the report and may be confusing to the reader. While the terms are essentially synonymous, each is commonly used and understood separately as part of the vernaculars of different organizations, cultures, and groups. Therefore, we have purposely elected to use both terms in the text of this report; Box 1-3 includes specific definitions of each of them.

BACKGROUND

Traditional Approaches for Preventing and Controlling Animal Diseases

Historically, measures taken at the national level to prevent animal diseases began at the country's borders and focused inward. The federal agency charged with primary responsibility for overseeing disease initiatives for livestock and poultry has been the U.S. Department of Agriculture's Animal and Plant Health Inspection Service (USDA APHIS). The traditional mission of APHIS, "to protect American agriculture," was carried out by channeling resources into three principal areas: adoption of quarantine measures to protect on-farm commodity production, implementation of emergency actions from the incursion of exotic diseases or related pests, and treatment to control or eliminate diseases and related pests.

The overall credibility of such efforts, with both domestic producers and other countries, hinged on the ability of APHIS, working with state counterparts, animal health professionals, and laboratories, to establish effective diagnostic systems, carry out continual inspection and surveillance, and respond to unforeseen emergencies from disease incursions. Ports of entry, inspection, and surveillance systems were established to prevent the introduction and spread of unwanted livestock and poultry diseases.

In addition to building and maintaining response capabilities, eradication programs were carried out for specific diseases such as foot-and-mouth disease, hog cholera, tuberculosis, brucellosis, and screw worm. Such programs required skilled expertise in specific disciplines such as veterinary medicine and were very labor-intensive. The disease profiles were generally well understood but required many years for complete eradication, which in some cases has still not been accomplished. Pockets of selected diseases, such as brucellosis and tuberculosis, still remain. Disease eradication campaigns required large financial outlays over a number of years, but to reduce exposure and protect growing export markets, campaigns for selected diseases such as foot-and-mouth or pests such as screw worm were funded and jointly carried out in other countries.

For many decades, this traditional approach looking at only livestock or poultry diseases—built on the presence or absence of specific diseases and supported by systems of inspection, diagnosis, and response—has served as the backbone of efforts to protect primary production. With the exception of ongoing screw worm eradication efforts in Central America and a relatively small workforce stationed across the world, direct financial support and involvement typical of the large international eradication campaigns of the past have been substantially scaled back or curtailed.[1]

While the animal health framework has been able to respond to a range of demands and challenges in the past, the pressures on animal health are increasing. In addition, there is a new recognition that wildlife and companion animals are playing a more important role in our lives and in the transmission of diseases that will further challenge the framework.

Changes Affecting the Animal Health Framework

Challenges to the animal health framework are, in part, the result of the transformation, restructuring, and fundamental changes in agriculture itself. A rapid consolidation in the agricultural sector is evident: the total number of U.S. farms declined from nearly 7 million at the beginning of the last century to less than 2 million today, while the average size of beef, dairy, pork, and poultry operations increased substantially (USDA, 2002c). At the same time, new sectors of animal agriculture (such as aquaculture and organically produced animals) are emerging. The last decade has also seen a marked increase in the number of fish raised in aquaculture facilities. In 2002, an estimated 6,653 farms (57 percent of which were less than 50 acres in size) produced 867 million pounds of aquaculture products (USDA, 2002c). Urban growth is increasingly encroaching on rural and wildlife environments, bringing populations of humans and animals, both farmed and wild, into closer and more frequent contact. Ecological systems and cultures that once were distinct are increasingly blurred at their borders, creating new opportunities for disease transmission and exchange at their interfaces. Furthermore, globalization has led to increased movement of products and animals worldwide and, by association, increased the potential for transmission of diseases and related pests between countries, while the genetic homogeneity of production animals may increase the vulnerability of the food

[1]APHIS is active in several disease and pest campaigns such as hog cholera in Hispanola and screw worm in Jamaica, but the scope, size, and role of APHIS, in comparison to campaigns of years past, is greatly reduced.

supply to catastrophic losses of disease. It is no longer sufficient to focus only on livestock or poultry, because many other species can be affected. Increasing population densities of domestic and some wild animal species have magnified the impact of infectious diseases.

One of animal agriculture's most difficult shifts is likely to be a move from its past independence to a new interdependence. This shift will be characterized by a profound interconnectedness in which producers and agriculturalists will be influenced and impacted by environmentalists, animal welfare activists, public health officials, agribusinesses, trade officials, and consumers. Today's agriculture is neither traditional nor in control of its own destiny. Increasingly, new sectors and special interest groups that have not been aligned or involved with agriculture in the past are now helping to shape its future. Policymakers, business leaders, and politicians—who are progressively more urban in thought and locale—envision animal agriculture and its future with a very different perspective. The complex, global, and intertwined world of contemporary agriculture will most certainly change the entire framework of animal health, how it operates, with whom it partners, and how it relates to a world that is rapidly closing in around it.

At the same time, an examination of the animal health system must assess its ability to coordinate and integrate actions within a larger group of participants who bring new perspectives and expectations for consideration. Inherent in this broadening scope and scale is the reality of emerging infectious diseases, new zoonoses, food safety problems, and the unfortunate reality that the intentional introduction of animal diseases could result in a cascading effect of potentially catastrophic consequences. The framework to address these diseases has increased in importance, complexity, and visibility. Certainly, the framework needs to be commensurate with, and responsive to, the profound forces and changes driving its future.

Relationships of Animal Diseases to Sectors beyond Production

The importance of animal diseases and related programs on domestic consumption, production, and trade has been well recognized, yet the impacts can include other dimensions and sectors such as food security, public health, market and product accessibility, economic viability, tourism, biotechnology, bioterrorism, and the environment.

The U.S. Agency for International Development (USAID) defines food security as "When all people at all times have both physical and economic access to sufficient food to meet their dietary needs for a productive and healthy life" (USAID, 1992). For the period 1999 to 2009, worldwide population is estimated to grow by 30 percent to 7.5 billion (IFPRI, 1999), and

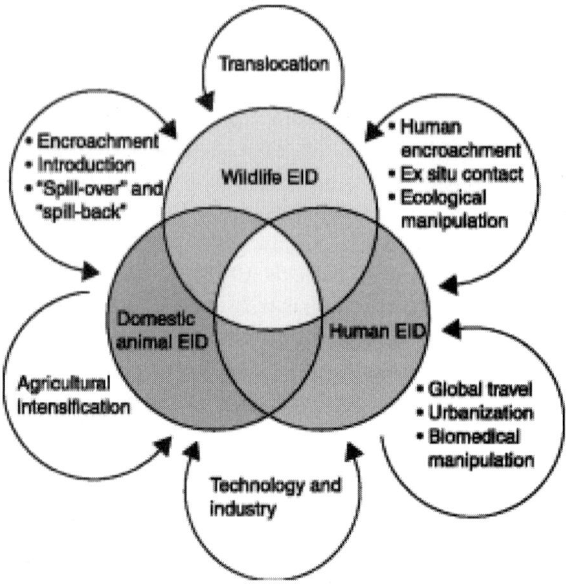

FIGURE 1-1 Interactions of emerging infectious diseases (EIDs) with a continuum that includes wildlife, domestic animals, and human populations. Few diseases affect exclusively any one group, and the complex relations among host populations set the scene for disease emergence. Examples of EIDs that overlap these categories include Lyme disease (wildlife to domestic animals and humans) and rabies (all three categories). Companion animals are categorized in the domestic animal section of the continuum. Reprinted with permission from Daszak et al., *Science* 287:443-449 (2000). Copyright 2004, AAAS.

between 1997 and 2020, it is estimated that worldwide demand for meat may increase 55 percent (IFPRI, 2003). Increasing trade and the need to ensure food security underscore the dual objective in carrying out agricultural animal disease prevention and control initiatives: to evaluate imports so that domestic production is not put at greater risk of disease, and to ensure that such initiatives do not become a bottleneck for the flow of products, limit access to products, or hamper food security. Traditional programs have understood the first objective well, but the second role has received little attention.

The Interaction among Domestic Animals, Wildlife, and Humans

As illustrated in Figure 1-1, the animal health framework must now deal with a continuum of host-parasite relationships involving public health, companion animals, wildlife, ecosystems, and food systems in an increasingly complex, global context.

Wildlife

Historically the attention paid to wildlife disease by regulatory agencies, wildlife advocacy groups, and educational and research institutions has been limited in comparison to that for domestic animals. However, the greatly increased contact among wildlife, domestic animals, and people associated with globalization and societal incursions into wildlife habitat has increased the opportunity for the transmission of pathogens shared within and among these domains. Increased population densities of people, domestic animals, and some wildlife species favored by human societal impacts on the environment, can make the situation even more precarious, because of the increased opportunity for transmission of infectious agents. For instance, expanding and increasingly dense poultry populations increase their chances of exposure to strains of avian influenza virus from wild waterfowl; expanding deer populations can support increased populations of deer ticks, which may spread Lyme disease where the causative agent *(Borrelia bergdorferi)* is endemic in the rodent population; and the periurban growth in raccoon populations increases the risk of rabies for humans and domestic animals. It is recognized that prevention and/or elimination of these disease interfaces is problematic, especially when steps to reduce wildlife populations can be difficult from both biological and sociological perspectives.

The host-parasite ecological continuum—described by Daszak et al. (2000) as a continuum where disease boundaries among wildlife, domestic animal, and human populations are increasingly blurred—provides a useful context in which to illustrate the importance of wildlife diseases and the pathogens they harbor in today's world (see Figure 1-1). The framework for dealing with animal diseases in this continuum must involve a wide range of government agencies and stakeholders that have overlapping interests in disease involving wildlife, as well as people and domestic animals. Providing the administrative means for effective coordination among the elements of such a framework is a formidable challenge.

West Nile virus (WNV) in wild birds, SARS in civets, avian influenza in wild migratory water fowl, plague (*Yersinia pestis*) in wild rodents, and rabies in raccoons, skunks, and bats are but a few examples of pathogens harbored in wildlife that can be transmitted to humans and/or domestic animals (see Chapter 3 for case studies). Conversely, pathogens are also being transmitted from domestic animals to wildlife populations; for example, tuberculosis to deer, elk, and bison; brucellosis from elk and bison to cattle; and *Mannheimia* pneumonia (pasteurellosis) in bighorn and domestic sheep (USAHA, 2003b). Tuberculosis in Michigan may have spilled back again from deer to infect cattle. In Yellowstone National Park, the presence of brucellosis in elk and bison is considered a potential threat to domesticated cattle grazing at the park boundaries (Dobson and Meagher, 1996). Zoo animals and exotic companion animals are also part of the disease continuum. For example, 58 zoo animals of 17 species in the United Kingdom acquired scrapie-like spongiform encephalopathies thought to result from exposure to feed contaminated by the BSE agent (Collinge et al, 1996; Kirkwood and Cunningham, 1994). Companion animal interaction with wildlife can also lead to transmission of pathogens, e.g., canine distemper to blackfooted ferrets (Thorne and Williams, 1988); feline leukemia virus to Florida panthers (JAVMA News, 2004); skunk rabies to cats; and a variety of parasitic helminths and protozoa to domestic livestock and people (Dubey and Lindsay, 1996; Waldner et al., 1999).

Translocation of both indigenous and exotic wildlife continues to be a major anthropogenic cause for the spread and impact of pathogens harbored in wildlife. This is a surprising circumstance because the danger inherent in such activity has been recognized for a long time. The dissemination of chronic wasting disease (CWD) and monkeypox are cases in point (see Chapter 3). Migratory wildlife can play a major role in spreading infectious agents as exemplified by rabies virus in mammals and avian influenza (AI) virus and Newcastle disease virus (NDV) in waterfowl.

During its deliberations the committee came to appreciate more fully the rapidly growing importance of wildlife as a source of infectious disease for people and domestic animals. While the committee had the expertise to assess the threat posed to humans and domestic animals by potential pathogens harbored in wildlife, it did not have members with sufficient knowledge of all the government agencies dealing with wildlife diseases to undertake a detailed assessment of their effectiveness. The federal and state responsibilities pertaining to all aspects of animal health and disease, including both domestic and wild terrestrial and aquatic animals, span such a wide array of government agencies that a detailed analysis of their functional relationships was beyond the practical capabilities of the committee. (For a chart of these agencies, see Figure 2-1 in Chapter 2.)

The committee recognized the importance of fish disease in both aquaculture (including coastal marine elements) and wildlife fishery. Aquaculture is the most rapidly growing livestock industry in the United States. Maintaining animal health is no less an issue for fish stock than for terrestrial animals. For example, recently the industry has had to cope with the introduction of exotic diseases such as infectious salmon anemia and spring viremia of carp in 2001 and 2002, respectively (USDA APHIS-VS, 2002; USDA, 2004c). Disease-related services at the federal level are fragmented among the U.S. Department of Agriculture's Animal and Plant Health Inspection Service (USDA APHIS), the U.S. Department of the Interior's Fish and Wildlife Service (DOI FWS), and the U.S. Department of Commerce's National Oceanographic and Atmospheric Administration (DOC NOAA). The need for coordination among these agencies is being addressed at present by a task force that is actively developing a National Aquatic Animal Health Plan under the auspices of the Joint Subcommittee on Aquaculture and the National Science and Technology Council Committee on Science, to be completed in 2006 (USDA, 2004a).

The Convergence of Human and Animal Health

Diseases found in humans have always been intensely affected by human-animal interactions. In fact, it is accepted that many infections of humans have origins in common with animals (Childs and Strickland, 2000). Although there are some diseases that are transmitted between humans only (for example, syphilis), a large number of domestic animal diseases are shared with humans—60 percent of the 1,415 diseases found in humans are zoonotic, and most are "multispecies" for domestic animal diseases (Cleveland et al., 2001).

With the development of agriculture approximately 10,000 years ago and the domestication of dogs and later livestock, animals became a more prominent part of our lives. Although there is good evidence to suggest that the advent of agriculture brought with it the phenomenon of zoonotic diseases, a new era of emerging and reemerging zoonotic diseases appeared to begin several decades ago. Since the mid-1970s, approximately 75 percent of new emerging infectious diseases of humans have been caused by zoonotic pathogens. Similar to the time of animal domestication, which triggered the first zoonoses era a number of millennia ago, a group of factors and driving forces have created a special environment responsible for the dramatic upsurge of zoonoses today.

The transmission of animal diseases to humans most often occurs via food through poor hygiene or improper handling of animal products. Organisms that cause zoonoses (such as bacteria, viruses, parasites, and protozoa) can also be transmitted via air, water, and vectors such as mosquitoes. In the field of emerging diseases, vector-borne and rodent-borne diseases are especially notable since they remain major causes of morbidity and mortality in humans in the tropical world and include a large proportion of the newly emerged diseases (IOM, 2003). The spectrum of vector-borne diseases are from animal-to-animal (bluetongue), animal-to-human (WNV), or human-to-human (dengue). It has been estimated that one tick-borne disease has emerged in the United States every decade for the past 100 years (IOM, 2003).

Some scientists argue that, of the more than 30 emerging diseases recognized since 1970, none are truly "new" but instead only newly spread to the human population (Saritelli, 2001). Today, new human behaviors create new risks for animal disease transmission to humans; for example, the feeding of animal by-products to cattle (which are herbivores) allowed BSE to emerge and spread to beef-eating humans, and the xenotransplantation of animal organs into humans raises concerns. The industrialization of food production, while making standardization of food easier, has created opportunities for large-scale microbial colonization in food animals and subsequent large-scale food contamination with pathogens, such as salmonella and *E. coli* O157:H7. Even when concerns are raised prior to transmission, action is not often taken. For example, it could have been predicted before 2003 that the importation of African rodents, including giant Gambian pouched rats, from areas where rodents are known to carry monkeypox would introduce monkeypox virus into the United States and potentially spread to humans. Today these pouched rats are being trained to detect landmines (Wines, 2004). If these trained rats are sent to various regions of the world to assist with landmine removal but precautions are not taken to ensure that they do not carry the monkeypox virus, monkeypox may spread to rodents and humans elsewhere in the world.

The confluence of people, animals, and animal products within today's dynamic international context is unprecedented, and we continue to face new microbial threats, as evidence by a recent outbreak of *E. coli* O157:H7 in a group of children who were exposed in a petting zoo, a Marburg virus outbreak in Angola, and the monitoring of H5N1 avian influenza in Southeast Asia as a potential pandemic strain. These concurrent events underscore the importance of new scientific and programmatic partnerships between veterinarians and public health officials and should serve as an impetus for the animal health framework to ensure a new capacity and focus that will address emerging and reemerging zoonoses.

ORGANIZATION OF THE REPORT

The next three chapters of this report review, summarize, and evaluate the state and quality of the current system for safeguarding animal health and the potential for improved application of scientific knowledge and tools. Chapter 2 provides a general overview of the animal health framework, supplemented by Appendix C, which contains additional details on the existing federal system for addressing animal diseases. Chapter 2 is also intended to supply the context for the entire three-phased initiative, and as such, includes issues relevant to surveillance, monitoring, response, and recovery. Chapter 3 focuses on prevention, detection, and diagnosis, exploring past experiences with a number of specific animal diseases. Based on the analysis of that experience, Chapter 4 considers current capabilities and limitations of the existing animal health framework and identifies key gaps in our ability to prevent, detect, and diagnose animal diseases. Finally, Chapter 5 provides recommendations for strengthening the existing system and identifies opportunities and needs for front-line prevention, detection, and diagnosis.

2

State and Quality of the Current System

INTRODUCTION

The committee was charged to review, summarize, and evaluate the state and quality of the current animal health framework. This review is organized into the following categories:

- Components of the Animal Health Framework
- Technological Tools for Preventing, Detecting, and Diagnosing Animal Diseases
- Scientific Preparedness for Diagnosing Animal Diseases: Laboratory Capacity and Capability
- Animal Health Research
- International Issues
- Addressing Future Animal Disease Risks
- Education and Training
- Improving Awareness of the Economic, Social, and Human Health Effects of Animal Diseases.

COMPONENTS OF THE ANIMAL HEALTH FRAMEWORK

The animal health framework comprises organizations and participants in the public and private sectors directly responsible for maintaining the healthy status of all animals and those who are impacted by animal health or are influencers of forces affecting animal health. The essential components of the framework for addressing animal disease, beginning with the affected animal, are listed in Box 2-1.

BOX 2-1
Components of the Animal Health Framework

1. People on the front lines of the animal production unit, animal habitat, or companion animal owners
2. Veterinarians and other sources of professional advice and care for health-related issues (such as universities and diagnostic laboratories)
3. Federal, state, and local animal health and public health agencies
4. International collaborations among agencies, organizations, and governments
5. Supporting institutions, industries, and organizations (including educators and the public health and intelligence communities)

Front Lines

The front lines contain multifarious actors and components: from intensive, large-scale, highly technical food animal facilities, monitored by well-trained livestock managers and veterinarians, to disparate clusters of companion animals within individual homes observed with differing degrees of intensity by their owners, to wildlife populations without any kind of regular monitoring contact by humans. It is a sine qua non that the first signs of a disease outbreak are small abnormalities in behavior. The sooner a new disease is recognized, the greater the likelihood that it will be effectively controlled and cause minimal damage.

In this context, an effective framework for animal health is most highly developed for agricultural animals. In today's livestock industry, producers are encouraged to adopt herd health programs and focus on prevention rather than dealing with case-by-case problems (Gary Weber, National Cattlemen's Beef Association, presentation to committee, April 6, 2004). As front-line responders, animal attendants and caretakers may have variable levels of training and motivation for recognizing and reporting abnormalities and sounding an alert when abnormalities are noted.

Farm animals are also raised by individual "hobbyists" who might lack the training of paid animal attendants but who potentially have the luxury to be more observant of their animals than do large-scale animal producers. They might also have expendable income with which to seek out veterinary services when needed. Because the number of hobbyists is growing, a better picture of the animal care practices of this community is needed to evaluate the knowledge of this group of owners and their likely motivation for reporting suspected disease outbreaks.

For companion animals and wildlife, the situation is even more uncertain. With the exception of some large charismatic and commercially viable species, there is little economic incentive to survey animal health, and in some cases, an absence of financially remunerated attendants responsible for monitoring husbandry. In these cases, recognition of a disease abnormality by people not associated with the immediate habitat is due to both diligence and chance. An astute owner may seek advice on first blush of a disorder in a companion animal, or alternatively, a group of companion animals may become quite ill prior to any abnormality being reported outside of the immediate surrounding. For wildlife, especially wildlife outside the oversight of zoo veterinarians and handlers, the situation can be even more uneven. For large and charismatic species (e.g., chimpanzees, giraffes, dolphins), detection of anomalies may occur at the early stages of disease development; however, with the majority of wild species (e.g., rodents, small birds, reptiles), disease may become widespread before it is recognized by people not associated with the immediate habitat.

Veterinary Medical Profession

The goals of the veterinary profession in the United States, as embodied in the oath taken by its members, are to protect animal health, relieve animal suffering, conserve animal resources, promote public health, and advance medical knowledge. In 1994, 56,000 veterinarians were active in the profession. In 2004, that number had grown to 65,000, a 16 percent increase. The profession is expected to grow another 25 percent in the next 10 years. The Bureau of Labor Statistics expects 28,000 job openings by 2012 due to growth and net replacements—a turnover of nearly 38 percent (AAVMC, 2004). Present employment of veterinarians is described in Table 2-1. Each state is responsible for licensing veterinarians and for regulating private veterinary practice (AVMA, 2004a).

The American Veterinary Medical Association (AVMA), established in 1863, serves as the lead professional body for veterinarians in the United States. It is an organization largely driven by private practitioners, the majority of whom are in companion animal practice and AVMA's primary activities are a reflection of the membership. It has a significant influence on veterinary education through its accreditation process administered by the Council on Education (COE). The AVMA also promulgates many and varied policy statements and guidelines that bear on animal health and welfare and on public health.

The United States Animal Health Association (USAHA) is another key organization dealing with agricultural animal health and disease issues. USAHA works with state and federal governments, universities, veterinarians, livestock producers, national livestock and poultry organizations, research scientists, the extension service, and seven foreign countries to control livestock diseases in the United States (USAHA, 2005). This coalition of government, academic, and industry animal health professionals has operated for more than 100 years and serves to discuss prominent issues and deliver resolutions to appropriate organizations and government for consideration.

The nature of veterinary employment is changing (Table 2-1). Over the past 15 years, there has been a 35 percent increase in the number of veterinarians engaged in small animal practice, a 13 percent decrease in the number of veterinarians in food-animal and mixed practice, and a 47 percent decrease in the number of veterinarians in public practice (i.e., government employment). Currently over half the profession is employed in small animal practice and only about 16 percent serves the livestock industry and food system, assuming that all the work of government employees is related to this domain (AVMA, 2005b).

The veterinary medical profession and its branches have been the subject of several in-depth assessments over the past 35 years (NRC, 1972, 1982, 2004b; Pritchard, 1988; Brown and Silverman, 1999). The KPMG megastudy conducted by Brown and Silverman (1999), entitled *The Current and Future Market for Veterinarians and Veterinary Medical Services in the United States*, examined the profession's income disparities, the increasing demand of services in new areas, and the critical shortage of trained professionals, and concluded that a series of strategic and substantive changes are needed in the veterinary profession to meet evolving societal needs and demands. One of the most comprehensive reviews, the Pew Veterinary Education Program, concluded: "Veterinary medicine is being threatened as never before by powerful forces of change in society, rapid advances in science and technology, and by the changing needs and expectations of almost every constituency it serves. Decisive steps must be taken at this time to make corrections in the way that the profession is trying to fulfill its responsibilities, to bring them more in line with the changing needs of society. Although it can not yet be defined as a crisis, the veterinary profession is not adapting rapidly enough to changing needs and is encountering substantial problems" (Pritchard, 1988). More recently, a 2004 National Research Council report on the veterinary medical profession found, among other key factors negatively impacting the supply of comparative medicine veterinarians, a lack of qualified applicants for all types of postgraduate training programs and the lack of commitment by veterinary medical schools and institutions that offer postgraduate training programs to prepare and train veterinary students and postgraduates for veterinary careers other than private clinical practice

TABLE 2-1 Employment of U.S. Veterinarians Who Are AVMA Members

	2004		1986	
Private Clinical Practice	Number	Percentage	Number	Percentage
Large animal exclusive	1,887	4.0	1,936	5.7
Large animal predominant	2,596	5.4	4,570	13.5
Mixed animal	3,868	8.2	3,397	10.1
Small animal predominant	5,507	11.7	4,722	14.0
Small animal exclusive	29,951	63.4	17,276	51.1
Equine	2,257	4.8	1,888	5.6
Other	1,198	2.5		
Subtotal	**47,264**	**100**	**33,789**	**100**
	2004		1986	
Public and Corporate Employment	Number	Percentage	Number	Percentage
College or university	3,961	46.7	3,713	39.5
Federal government	641	7.6	2,212	23.5
State or local government	542	6.4	756	8.0
Uniformed services	474	5.6	586	6.2
Industrial	1,566	18.5	2,128	22.7
Other	1,294	15.2		
Subtotal	**8,478**	**100**	**9,395**	**100**
Grand Total	**64,867**		**43,184**	

SOURCE: Pritchard, 1988; AVMA, 2005b.

(NRC, 2004b). While it is too early to tell whether the recommendations from the 2004 NRC report have had an effect, the employment demographics of veterinarians over the last 15 years (Table 2-1) suggest that many of the Pew report recommendations have not been realized, due largely to the limited amount of funding provided and the complete lack of follow-up and continuity.

Private Veterinarians

Veterinarians in private practices, generally supported by veterinary technicians, are among the front-line health professionals dealing with animal disease. They constitute about 80 percent of the veterinary workforce (ca. 47,000, as shown in Table 2-1). Fewer than 10,000 derive a significant portion of their income from food-animal practice, and the number is declining (AVMA, 2005b). Rural demographic changes, intensification and specialization in the livestock industry, lifestyle issues, veterinary college entrance selection, and perhaps shifts in gender balance have led to circumstances where fewer veterinary graduates opt for careers in rural food animal practice (AVMA, 2004b).

Veterinarians working in small animal and exotic practice can also play key roles in the detection of emerging disease problems. Examples of successful recognition of early incursions include the diagnosis of West Nile virus by a veterinary pathologist at the Bronx Zoo and screw worm incursions halted by small animal and equine practitioners in two different states (Nolen, 1999; Thurmond and Brown, 2002).

Federal and State Animal Health Agencies

Federal Animal Health Agencies

This section briefly summarizes the legal authorities and functions of the federal government for preventing, detecting, and diagnosing animal diseases. Appendix C contains a more detailed summary prepared by Nga L. Tran, entitled "Existing Federal System for Addressing Animal Diseases." Figure 2-1 illustrates the large number of federal entities involved in addressing animal health issues. International, state, and private entities involved in animal health issues are not included in Figure 2-1.

The USDA Animal and Plant Health Inspection Service (APHIS) plays the lead role in protecting the health of domestic animals. Within APHIS, the majority of the responsibility to protect animal health resides in Veterinary Services (VS). The USDA's programs addressing animal health cover a wide range of functions, including deterrence (the elimination or reduction of factors conducive to the potential import, transport, or transmission of disease from suspected sources of pathogens) and prevention, detection and diagnosis, monitoring and surveillance, emergency response, research, education and training, and communication (see Table C-3). A summary of deterrence and prevention efforts as they relate to reducing a potential threat before it reaches U.S. borders are described later in this chapter in the section on International Issues.

The APHIS-VS division shares responsibility for some animal health issues with the Food and Drug Administration's (FDA) Center for Veterinary Medicine (CVM). The CVM regulates and approves the manufacture and distribution of food additives and drugs that will be given to animals. APHIS-VS's Center for Veterinary Biologics (CVB) regulates veterinary biologics, including vaccines, bacterins, antisera, and diagnostic kits that are used to prevent, treat, or diagnose animal diseases and ensure that these products are pure, safe, potent, and effective,

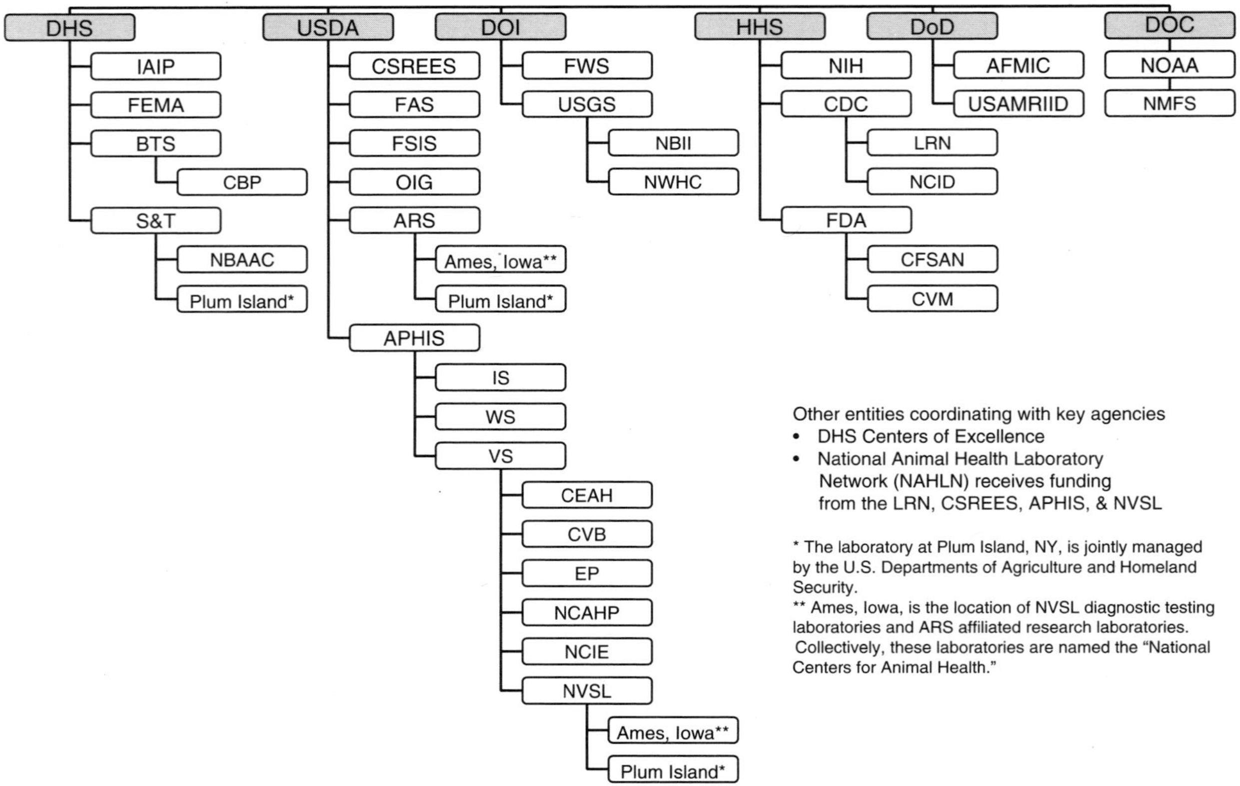

FIGURE 2-1 Key federal agencies addressing animal diseases.

Glossary of Acronyms and Abbreviations for Figure 2-1

AFMIC – Armed Forces Medical Intelligence Center
APHIS – Animal and Plant Health Inspection Service
ARS – Agricultural Research Service
BTS – Border and Transportation Security
CBP – Customs and Border Protection Bureau
CDC – Centers for Disease Control and Prevention
CEAH – Centers for Epidemiology and Animal Health
CFSAN – Center for Food Safety and Applied Nutrition
CSREES – Cooperative State Research, Education, and Extension Service
CVB – Center for Veterinary Biologics
CVM – Center for Veterinary Medicine
DHS – U.S. Department of Homeland Security
DOC – U.S. Department of Commerce
DoD - U.S. Department of Defense
DOI - U.S. Department of the Interior
EP – Emergency Programs
FAS – Foreign Agricultural Service
FDA – Food and Drug Administration
FEMA – Federal Emergency Management Agency
FSIS – Food Safety and Inspection Service
FWS – Fish and Wildlife Service Bureau
HHS – U.S. Department of Health and Human Services
IAIP – Information Analysis and Infrastructure Protection Directorate
IS – International Services
LRN – Laboratory Response Network
NBII – National Biological Information Infrastructure
NBAAC – National Biodefense Analysis Countermeasure Center
NCAHP – National Center for Animal Health Programs
NCID – National Center for Infectious Diseases
NCIE – National Center for Import and Export
NIH – National Institutes of Health
NMFS – National Marine Fisheries Service
NOAA – National Oceanic and Atmospheric Administration
NVSL – National Veterinary Services Laboratories
NWHC – National Wildlife Health Center
OIG – Office of the Inspector General
S&T – Science and Technology Directorate
USAMRIID – U.S. Army Medical Research Institute for Infectious Diseases
USDA – U.S. Department of Agriculture
USGS – U.S. Geological Survey
VS – Veterinary Services
WS – Wildlife Services

based on the Virus-Serum-Toxin Act (37 Stat. 832-833; as amended December 23, 1985, Pub. L. 99-198, 99 Stat. 1654-1655; 21 U.S.C. 151-159).

The APHIS-VS division also administers the National Veterinary Accreditation Program (NVAP). This voluntary program certifies private veterinary practitioners to work cooperatively with federal and state animal health officials. Nationally, more than 60,000 active accredited veterinarians are in the NVAP database. These veterinarians are instrumental in performing examinations and issuing health certificates critical to the safe movement of animals, assisting in disease eradication campaigns, and maintaining extensive animal disease detection and surveillance functions. NVAP work must be consistent with international requirements to safeguard animal health.

The USDA Food Safety and Inspection Service (FSIS) is responsible for ensuring the safe, wholesome, and correctly labeled and packaged commercial supply of meat, poultry, and egg products that move within interstate commerce, are imported into the United States, or exported to other countries. Over the years, FSIS has transitioned into a public health role and has especially focused on food safety and security. Through its inspection system, which involves inspection of individual animal carcasses at slaughter, FSIS plays an important disease detection function. For instance, FSIS assists APHIS in identifying tuberculous cattle carcasses for the national bovine TB eradication program. The FSIS inspection system is further enhanced through its use of toxicological, pathological, and microbiological analyses. In this capacity, the agency is able to help prevent the dissemination of pathogens and diseases to people and animals further along the commodity stream. FSIS employs approximately 7,600 inspectors and is the largest employer of veterinarians in the federal government.

The Fish and Wildlife Service (FWS) of the U.S. Department of the Interior (DOI) is responsible for the protection and enhancement of wildlife populations, safeguarding habitat for wildlife, including endangered species, and the inspection of wildlife shipments imported into the United States to ensure compliance with laws and treaties and detect illegal trade (FWS, 2001). DOI's National Wildlife Health Center (NWHC) was established in 1975 as a biomedical laboratory dedicated to assessing the impact of disease on wildlife and identifying the role of various pathogens contributing to wildlife losses (USGS, 2004). The center provides a multidisciplinary, integrated program of disease diagnosis, field investigation and disease management, research, and training. It also maintains extensive databases on disease findings in animals and on wildlife mortality events. Other DOI programs include the National Biological Information Infrastructure (NBII), a broad collaborative program providing increased access to data and information on the nation's biological resources.

HSPD-9

On January 30, 2004, the White House issued a policy directive, Homeland Security Presidential Directive-9 (HSPD-9), which makes the U.S. Department of Homeland Security responsible for coordinating federal programs aimed at protecting U.S. agriculture and food from diseases, pests, and toxins. Veterinary medicine is a critical component of HSPD-9, which significantly expands federal animal health-related initiatives. For instance, the policy calls for creation of a national stockpile of animal drugs and vaccines to respond to serious animal diseases; grants to veterinary colleges for expanding training in exotic animal diseases, epidemiology, and public health; and inclusion of veterinary diagnostic laboratories in national networks of federal and state laboratories (The White House, 2004).

Over the course of 2004, federal response to HSPD-9 and related Homeland Security presidential directives was initiated and included in a USDA Agriculture Emergency Response Training session targeting APHIS animal health personnel and a scientific conference targeting development and use of rapid detection technologies. In January 2005, the Department of Homeland Security released its National Response Plan in response to HSPD-5 (Management of Domestic Incidents), which includes elements supportive of HSPD-9 efforts. The National Response Plan serves to "align federal coordination structures, capabilities, and resources into a unified, all-discipline, and all-hazards approach for incident management" (DHS, 2004d) and includes notation that annexes specific to food and agriculture will be published in subsequent versions of the plan.

State Animal Health Agencies

With few exceptions, states have the greatest responsibilities for animal health, whether for agricultural animals, companion animals, or wildlife. Local authorities will quickly become involved in an animal health emergency, but as soon as resources are overwhelmed, the state will assume responsibility. The federal government oversees issues involving foreign animal and programmatic diseases, veterinary biologics, and national identification and surveillance systems. It also monitors animals at U.S. borders, serves as a reference laboratory, and regulates imported and exported animals and animal products. Most all other animal health issues are dealt with at the state level or as a part of a cooperative state-federal program.

The state departments of agriculture play a vital role in the animal health framework. Through their departments of agriculture, each state assumes responsibility to provide services and regulations regarding the health of agricultural animals. States maintain a list of reportable diseases and require all veterinarians to report disease occurrences. State veterinarians spearhead and direct the efforts of state animal health officials who have intrastate authority for disease reporting, detection, and often, diagnosis. These same officials also serve as key cooperators with their federal government counterparts in the prevention, detection, and eradication of a number of foreign and domestic diseases associated with national animal disease programs. In addition to state veterinarians, a large majority of states also have state public health veterinarians, whose positions and offices are usually associated with departments of public or community health. These officials are responsible for dealing with zoonoses and many other dimensions of veterinary and human public health. State agencies license veterinarians, regulate the intrastate movement of animals, organize emergency response, and are responsible for wildlife. States typically provide regulatory, laboratory, epidemiological, and programmatic support to their livestock, companion animal, and wildlife industries by working through veterinary practitioners, directly with producers, with relevant industries, and with local and federal animal and public health agencies.

A major contribution of the states is the maintenance of animal health diagnostic laboratories. In most states, animal health diagnostic laboratories are associated with state departments of agriculture and, depending on the state, are located at veterinary colleges, land grant university departments of veterinary science, or state agencies for public health. Some states have multiple laboratories. These facilities handle or forward the majority of specimens for diagnosis and monitoring of disease. Private laboratories also play an increasing role in the diagnosis of animal diseases, especially for companion animal species. More information about diagnostic laboratories is described in the section of this chapter entitled "Scientific Preparedness for Diagnosing Animal Diseases: Laboratory Capacity and Capability."

International Organizations

Many international organizations are involved with issues related to animal disease. Given the increasingly global nature of disease outbreaks, these agencies, the most important of which are highlighted here, play a key role with respect to the animal health framework in the United States. The agencies involved in prevention, detection, and diagnosis of animal diseases consist of several multilateral groups that have different mandates and functions but do not have national regulatory authority.

Nevertheless, as a member of many of these international organizations, the United States is committed to the responsibilities of membership. For example, in the case of the World Trade Organization (WTO), member countries are obligated to bring national laws in conformity with the WTO agreements and adhere to the dispute resolution procedures and outcomes.

The WTO deals with the global rules of trade between nations. Its overriding objective is to help trade flow smoothly, freely, fairly, and predictably. It achieves these objectives by administering trade agreements, acting as a forum for trade negotiations, settling trade disputes, and reviewing national trade policies. The WTO has nearly 150 member countries, which account for over 97 percent of world trade, and approximately 30 other countries are currently negotiating membership. Decisions are made by a consensus of the member countries. Agreements are the legal ground rules for international commerce and are essentially contracts, guaranteeing member countries important trade rights. They also bind governments to keep their trade policies within agreed limits to the benefit of all. While the agreements are negotiated and signed by governments, their true intent is to help the producers of goods and services, exporters, and importers conduct their business and to improve the welfare of the peoples of the member countries.

Critically important for the animal health framework is the Agreement on the Application of Sanitary and Phytosanitary Measures (SPS Agreement), which concerns the application of food safety and animal and plant health standards while allowing countries to set their own science-based standards. Member countries are encouraged to use international standards, guidelines, and recommendations where they exist: "The basic aim of the SPS Agreement is to maintain the sovereign right of any government to provide the level of health protection it deems appropriate, but to ensure that these sovereign rights are not misused for protectionist purposes and do not result in unnecessary barriers to international trade. The standards are developed by leading scientists in the field and governmental experts on health protection and are subject to international scrutiny and review" (WTO, 1998). For example, members may set more stringent standards if there is scientific justification or if it is based on an appropriate assessment of risks and the approach is not arbitrary.

In establishing the WTO/SPS Agreement, three international standard setting bodies were specifically referenced: the World Organization for Animal Health (OIE) for animal health and food safety issues of animal production, the International Plant Protection Convention for plant health, and the Codex Alimentarius for food safety. Of these so-called "three sisters," the OIE is the most relevant for monitoring animal health. The importance of the OIE in the animal health framework is to promote transparency in the global animal health situation through the collection, analysis, and dissemination of disease and health information; to encourage a coordinated approach to disease outbreaks; to safeguard world trade through animal health standards; to help define and support animal welfare and animal production food safety initiatives; and to improve national veterinary services through the determination of standards and levels of performance.

The OIE is an intergovernmental organization created in 1924 with 28 original member countries; it now has 167 member countries. Through the efforts of scientific commissions and participation of member countries, the OIE determines, revises, and publishes standards contained in the *Terrestrial Animal Health Code and Manual* and the *Aquatic Animal Health Code and Manual*. The OIE also collects and disseminates information on animal diseases, including changes in disease status and programs. The OIE has established 157 reference laboratories in 30 countries that are able to diagnose over 80 diseases and address related topics. The OIE also has 15 collaborating centers in 7 countries.

The World Health Organization (WHO), the program for food and agriculture within the International Atomic Energy Agency (IAEA), and the Food and Agriculture Organization (FAO) are branches of the United Nations. In terms of the animal health framework, the FAO focuses on food and animal health in developing countries. FAO activities include emerging and "transboundary" disease problems, i.e., those diseases that move with few barriers from one country to another and significantly hamper animal protein production and trade. Other forms of FAO technical assistance are technical advice, professional information, fielding of experts and consultants, provision of technical documentation, training, and preparation and execution of field projects in close cooperation with animal health services of member countries.

The IAEA program for food and agriculture contains a subprogram on animal health and disease, which is carried out in collaboration with the FAO. The subprogram promotes improved disease management through the application of nuclear and related biotechnologies. In this regard, much effort is focused on diagnostic and surveillance methods and strategies for priority livestock and poultry diseases in developing countries such as foot–and-mouth, exotic Newcastle, and African swine fever.

The WHO deals with diseases affecting humans, including zoonotic diseases. It contributes to animal health wherever human health is affected on an international scale. This regionalized organization has individual country, regional, and worldwide programs and responsibilities. The Veterinary Public Health (VPH) section, which deals with zoonoses and food hygiene, has access to the expertise of the many specialized WHO agencies.

The WHO regional office for the Americas is the Pan American Health Organization (PAHO), whose mission "is to strengthen national and local health systems and improve the health of the peoples of the Americas, in collaboration with Ministries of Health, other government and international agencies, nongovernmental organizations, universities, social security agencies, community groups, and many others." The PAHO assists the countries of Latin America and the Caribbean in dealing with health issues through their scientific and technical experts located in the United States, country offices, and scientific centers. For example, PANAFTOSA, a regional laboratory located in Brazil, was originally established to provide diagnoses of specific diseases, such as foot-and-mouth disease. The PAHO also provides support for disease eradication efforts in infected regions and neighboring countries.

Another organization operating in the Americas is the Inter-American Institute for Cooperation in Agriculture (IICA). IICA supports national veterinary services' efforts to: (1) develop regulatory mechanisms, science-based technical capacity, and sustainable institutional infrastructures; (2) apply the provisions of the Sanitary and Phytosanitary Agreements of the WTO as well as the decisions taken in the international reference organizations: OIE, IPPC, and Codex Alimentarius; and (3) assist countries with early recognition of emerging diseases and issues.

The International Regional Organization on Agriculture Health (OIRSA) works with the seven countries of Central America, Mexico, and the Dominican Republic (OIRSA, 2004). OIRSA provides support for the modernization of national services and related sanitary actions. Specific actions follow the disciplines outlined in the WTO/SPS Agreement and include harmonization, risk analysis, equivalence, and regionalization. It also seeks to strengthen inspection and quarantine control related especially to interregional trade and programs of prevention and control assistance with the harmonization of norms, risk analysis methodologies, surveillance, inspection controls, and support for disease eradication.

A discussion of the role of international developmental agencies, such as the U.S. Agency for International Development (USAID), foundations, nonprofit organizations, and regional banks, is beyond the scope of this report. Worth mentioning because of its relation to the WTO/SPS Agreement is the Standards Trade Development Facility (STDF), which is coordinated by the WTO/SPS Secretariat to assist countries in improving their sanitary status. Initial funding for the STDF was provided by the World Bank, one of the world's largest sources of development assistance, providing low-interest loans, grants, and interest-free credit to governments in developing countries for infrastructure improvements. The STDF is a global program providing technical assistance and capacity building to developing countries in implementing the measures contained in the SPS

Agreement. The STDF is both a financing and coordinating mechanism working with countries to improve their sanitary status and thus comply with and benefit from the SPS Agreement (STDF, 2004).

The subject of the role of international organizations in helping to secure animal health in the United States and globally is one that merits attention in greater detail in future examinations of the animal health framework.

Supporting Institutions, Industries, and Organizations

Supporting institutions, industries, and organizations also play a role in preventing, detecting, and diagnosing animal diseases. These include educational and research institutions, professional societies, and animal commodity groups. An in-depth examination of all of these entities is not presented here, but like international organizations, they play a role that should be examined more closely in future analyses.

TECHNOLOGICAL TOOLS FOR PREVENTING, DETECTING, AND DIAGNOSING ANIMAL DISEASES

Early detection, identification, and diagnosis are critical for limiting the extent of an animal disease outbreak and protecting the public from potential zoonotic disease exposures. Despite recent advances in technologies—including DNA-based techniques, novel sampling approaches, and more rapid, automated, and nonsubjective analytic tools—the classic laboratory techniques, which have changed little in the past 50 years, remain the most common means of identifying animal disease agents. Traditional bacteriological, fungal, viral, toxicological, and serological testing methods, though tried and true for several generations, require considerable investments in time (hours to days), extensive technical training and scientific judgment, and the prior recognition of a clinical problem in order to trigger testing of the animal or animal population. Sometimes the only material available for analysis is a dead or dying animal, which is frequently the case with wildlife and occasionally with food-animals. In those cases, immunohistopathological methods are needed to examine cellular changes caused by infections, such as the presence of inflammatory cells, viruses, or antibodies.

State-of-the-art scientific approaches that would enhance early detection and diagnosis of human disease are often developed by and for basic research and military applications and then rapidly adopted by first-responder and public health communities. The same technologies have been significantly slower to transition into the animal health arena. As noted in a prior NRC report, technological advances that speed and increase the reliability of the detection and diagnostic process have not been aggressively applied to agriculturally important pathogens, nor have they been inexpensive or field-deployable (NRC, 2003a). The same situation applies to virtually all animal disease agents, whether affecting wildlife, livestock, or companion animal species.

Transitional or applied research and federal funding sources to support the development, validation, and/or implementation of technological tools specifically for animal health applications are limited. Furthermore, economic incentives for the private sector do not traditionally support these development efforts. Federal and state laboratories across the country often have difficulties in acquiring advanced technologies such as robotics to increase the numbers of specimens or tests that could be processed with minimal human intervention (surge capacity), instrumental analyses (e.g., gas chromatography, mass spectrometry) for high resolution toxin and protein detection, and DNA-based tools that provide for rapid and sensitive agent detection or identification. There are multiple reasons for this situation, such as constraints on space, the lack of technical know-how or trained staff, or adequate numbers of samples to justify the acquisition of expensive equipment. Homeland security initiatives related to bioterror preparedness have improved both federal and state laboratory access to rapid DNA-based diagnostic tools such as realtime or quantitative polymerase chain reaction (PCR); however, as an industry, animal health lags years behind the military, first-responder, and public health communities in its implementation and use of advancing technologies.

In recent years, the movement of diagnostic assays out of the confines of the laboratory and into the field, closer to the source of the disease, has been made possible by scientific advances that provide the technology to shrink laboratory equipment by orders of magnitude (see Box 2-2 for "Examples of Evolving Technologies"). These technological advances, including miniaturization and microfluidics, allow use of increasingly smaller fluid volumes and microscopically thin equipment components and wiring, all of which allow chemical and physical reactions to occur faster and more uniformly. Devices that once required feet of laboratory space, relatively large volumes of clinical material, and large quantities of expensive assay components are now available in high-speed, portable, and in some cases hand-held forms. Sophisticated real-time PCR equipment, available just a decade ago only in high-tech laboratories, is now accessible to buyers in portable handheld or backpack versions targeting the first-responder and security communities. Access to size-reduced laboratory equipment has allowed fully functional mobile high-tech laboratories to be moved on site for immediate human health response capability, as seen in 2001 in Washington, D.C., during the anthrax letter scare and in 2002 in Salt Lake City for the Olympics. Similar portable laboratory approaches

BOX 2-2
Examples of Evolving Technologies That Enhance Prevention, Detection, and Diagnosis

PREVENTION
- Immune modulators and nonspecific stimulants
- Recombinant vaccines: "marker" vaccines with paired diagnostic assays
- Risk analysis tools: computational modeling

DETECTION AND DIAGNOSIS
- Polymerase chain reaction (PCR), real-time or qPCR
- Genomic sequence analysis, sequencing arrays
- Genechips, microarrays: liquid arrays
- Nanotechnology, biowires: computer chips for disease detection
- Environmental ("sniffer" technology) for detecting airborne pathogens
- Fully automated, integrated sampling, detection, and reporting systems for pathogens
- High-tech mobile laboratories

have been proposed for rapid on-site response to critical animal health issues, such as for a potential foot-and-mouth disease (FMD) outbreak in an area not immediately accessible to regional laboratory services.

Advances in biostabilization—essentially freeze-drying of unstable assay components—have also allowed movement of assays from the traditional laboratory to the field, as well as provided the foundation for fully automated on-site detection systems that continuously sample the air (termed "sniffers") and monitor high-risk environments. In 2000, author Richard Preston envisioned a portable environmental "sniffer" paired with PCR for genome-based detection of a bioterror agent, compacted into the size of a briefcase (Preston, 1997). Preston's novel, a fictionalized account of bioterrorism, was based on real-world developments that ultimately led to the U.S. Department of Homeland Security's BioWatch program. Since 2003, BioWatch and prototype commercial environmental sampling and test systems have continuously monitored selected public venues, including subways, banks, and post offices, for human biothreat agents (CDC, 2005; OSTP, 2005). Similar automated sampling and monitoring of high-risk animal environments for high-economic risk pathogens have not received equal attention.

The local public health community can choose to use a broad array of diagnostic tools ranging from simple rapid detection tests, such as at-home pregnancy kits, to more sophisticated assay formats, such as PCR. These formats are common, readily available, and standardized for the public health community, but not so for the community of veterinary laboratories and clinics. Nanotechnology, the ability to build at a scale of a billionth of a meter, is being described as the next technical revolution and may allow the development of electronic circuitry 1,000 times smaller than current microchips. Among others, one application for human and animal health includes the potential for embedded medical monitoring (chips inserted under the skin).

SCIENTIFIC PREPAREDNESS FOR DIAGNOSING ANIMAL DISEASES: LABORATORY CAPACITY AND CAPABILITY

Overview

The nation's animal health laboratory system is composed of federal, state, university, and commercial laboratories. The federal component is referred to as the National Veterinary Services Laboratories (NVSL). The NVSL, which is part of USDA-APHIS, provides diagnostic services through two testing facilities located in Ames, Iowa, and the Foreign Animal Disease Diagnostic Laboratory (FADDL) on Plum Island, New York. These laboratories perform the following functions: oversee and conduct laboratory testing in conjunction with federally mandated eradication programs for diseases such as brucellosis, pseudorabies, and tuberculosis; screen samples for the presence of exotic diseases at the request of federal and state regulatory staff; assist in investigating unusual agricultural animal disease occurrence in the United States; perform tests to meet animal export requirements; conduct testing for routine support of national and state animal health management; and serve as reference laboratories for certain infectious diseases (USDA, 2004b). However, the vast majority of routine diagnostic and animal health management analyses on domestic animals in the nation are conducted daily by state and university-affiliated veterinary diagnostic laboratories. The same is true for wildlife diseases. With few exceptions (for example, the U.S. Geological Survey's [USGS] National Wildlife Health Laboratory in Madison, Wisconsin; the Southeast Cooperative Wildlife Disease Study [SECWDS] in Athens, Georgia, which is a federal-state partnership; the FWS Forensics Laboratory in Ashland, Oregon; and scattered state wildlife agency-affiliated laboratories), routine investigation of wildlife diseases in the United States occurs in state/university diagnostic laboratories. Diagnostic work on zoo and exotic animal species is performed by laboratories associated with large municipal zoos and wildlife parks and by private zoo consultants who generally are board-certified pathologists. State, university, and commercial veterinary diagnostic laboratories also investigate diseases in zoo and exotic animals, often in response to requests from smaller zoos. Taxon advisory groups and species survival plans within the American Zoo and Aquarium Association may recommend particular laboratories for certain tests for the sake of consistency. A zoo and exotic animal laboratory network was recently established based largely on initial detection of West Nile virus infections in that environment (Nolen, 1999; Ludwig et al., 2002). This small but rather active network of diagnosticians and laboratories seeks to expand and integrate its activities with other veterinary and public health laboratory networks and is a key partner in detecting diseases that often emerge at the interface of exotic animals, domestic animals, and humans. Toward that goal, the International Species Information System is in the process of developing an information technology tool, termed the Zoological Information Management System (ZIMS), to replace the limited database software for zoo species.

The classical approaches of diagnosing diseases in agriculture cannot be transposed onto wildlife diseases. Whereas domestic animal populations can be manipulated and individual animals can be examined with relative ease to determine the prevalence or incidence, neither can be done with wildlife. Thus field work requires specialized expertise that is supported by research in techniques such as modeling, application of technologies such as GIS, and knowledge of pathology for field necropsy and/or collection of specimens. Diagnostic laboratory methodology, in contrast, is essentially the same as for domestic animals. Information about current wildlife disease issues is available through the Wildlife Diseases Information Node of the National Biological Information Infrastructure (NBII).

Laboratory Networks

National Animal Health Laboratory Network (NAHLN)

In June 2002, the Public Health Security and Bioterrorism Preparedness and Response Act of 2002 was signed into law. Section 335 authorizes the Secretary of Agriculture to develop an agricultural early warning surveillance system enhancing capacity and coordination between state veterinary diagnostic laboratories, federal and state facilities, and public health agencies, and provides authorization for Congress to appropriate funding to the National Animal Health Laboratory Network (NAHLN) (McElwain, 2003). The NAHLN addresses diagnostic needs for early diagnosis of exotic and emerging diseases and for routine animal disease surveillance, as well as diagnostic capacity for disease investigations, response and control, and eradication programs. The national laboratory concept was first developed in discussions between the American Association of Veterinary Laboratory Diagnosticians (AAVLD) and NVSL, resulting in a Memorandum of Understanding signed by NVSL and AAVLD in 2001 to "cooperatively improve animal health diagnostic services in the United States." In addition, the Safeguarding Review, commissioned by USDA to comprehensively review the federal system for safeguarding animal health in the United States, emphasized the need for a comprehensive and coordinated network (NASDARF, 2001). Initial support through cooperative agreements to 12 state diagnostic laboratories to establish the pilot NAHLN with NVSL was provided through emergency appropriation of Department of Defense (DoD) funds for Homeland Security through the USDA Cooperative State Research, Education, and Extension Service (CSREES) (USAHA, 2002).

The philosophy behind the design and implementation of NAHLN is that animal disease surveillance functions most effectively as a shared responsibility between federal and state animal health agencies. During a disease outbreak, state veterinary diagnostic laboratories would provide early diagnosis and significant surge capacity. State labs would assist and define herds for depopulation, delimit the extent of the outbreak, and conduct follow-up surveillance to determine a "disease-free status" (USDA, 2004d). The development of the NAHLN heralded a fundamental change in the animal health laboratory infrastructure in the United States. For the first time, the need for state laboratories to test for exotic pathogens was acknowledged. Perhaps more important, there was recognition of the responsibility of state laboratories to test in service to their stakeholders.

The main goals of the NAHLN are to expand detection and response measures for pathogens that threaten animal agriculture and bolster laboratory capability for select agents with support for personnel, equipment, testing, training, and information technology. Among the elements being implemented in the NAHLN are development and deployment of standard diagnostic approaches for identification of select agents; rapid diagnostic techniques, modern equipment, and experienced personnel trained in the detection of emergent, foreign, and bioterror agents; a national training program that ensures competency and consistency in diagnostic testing using new equipment and reagents; proficiency testing and quality assurance; and upgraded facilities that meet biocontainment requirements. The NAHLN is also developing an information technology tool to facilitate data sharing among animal health agencies through secure, automated, two-way communications to create a national repository for animal health data; bolstering cooperation and communication among animal health officials through maintenance of confidentiality of source data; and providing alerts at an appropriate response level (USAHA, 2002).

The pilot NAHLN involved 12 state/university diagnostic laboratories and was charged with developing capacity and surveillance programs for eight high-priority foreign animal diseases considered to be of bioterrorist threat (FMD, hog cholera, African swine fever, rinderpest, contagious bovine pleuropneumonia, lumpy skin disease, highly pathogenic influenza, exotic Newcastle disease). Other agents of interest, such as vesicular stomatitis, West Nile encephalitis, Rift Valley fever, Nipah encephalitis, Hendra encephalitis, scrapie, CWD, and BSE, will be added in the future (USAHA, 2002). NAHLN laboratory personnel have been trained in the standard nomenclature used in reporting laboratory results. It is anticipated that the number of NAHLN laboratories and the breadth of diseases covered in the NAHLN will increase significantly by FY 2009, creating a broader pool of expertise that can be tapped for surge testing capacity in an outbreak (USDA, 2004d). USDA APHIS and the CSREES recently agreed to expand the NAHLN definition to include, in addition to the original 12 laboratories in the pilot program, all federal and state laboratories currently contracting with USDA for scrapie, CWD, and AI/END surveillance testing. These additional laboratories, however, did not receive any additional funding beyond appropriate equipment to address infrastructure needs as discussed above.

The NAHLN administrative structure includes a national coordinator who reports to the director of the NVSL and a steering committee. In addition to the NVSL director and the NAHLN coordinator, the steering committee has representation from the state laboratories through AAVLD (two NAHLN laboratory directors, one non-NAHLN laboratory director, and the current president of AAVLD), a state veterinarian, and CSREES national program leader for homeland security.

Laboratory Response Network (LRN)

USDA APHIS and AAVLD are also partnering with the Centers for Disease Control and Prevention (CDC) to enlist state veterinary diagnostic laboratories into the CDC Laboratory Response Network (LRN). The LRN was established in 1999 to prepare the U.S. response to bioterrorism. The mission of LRN and its partners is to maintain an integrated national and international network of laboratories that is fully equipped to quickly respond to chemical or biological acts of terrorism, emerging infectious diseases, and public health threats and emergencies (Gilchrist, 2001). CDC runs the LRN program with direction and recommendations provided by the Association of Public Health Laboratories, the Federal Bureau of Investigation (Department of Justice), the AAVLD, the American Society for Microbiology, EPA, USDA, DoD, FDA, and DHS (CDC, 2005).

LRN is a consortium of 137 laboratories that can provide immediate and sustained laboratory testing and communication in the event of public health emergencies, particularly bioterrorism-related events. The network includes the following types of laboratories:

- Federal—laboratories at CDC, USDA, FDA, and other facilities run by the federal agencies.
- State and local public health—laboratories run by state and local departments of health.
- Military—laboratories operated by the DoD, including the U.S. Army Medical Research Institute for Infectious Diseases (USAMRIID) at Fort Detrick, Maryland.
- Food testing—FDA laboratories and others that are responsible for ensuring the safety of the food supply.
- Environmental—laboratories that are capable of testing water and other environmental samples.
- Veterinary—some LRN labs, such as those run by USDA and state veterinary diagnostic laboratories, that are responsible for animal testing. Some diseases can be shared by humans and animals, and animals often provide the first sign of disease outbreak.
- International—laboratories located in Canada, the United Kingdom, and Australia.

The LRN laboratories are designated as either national, reference, or sentinel. National laboratories (including those at the CDC and USAMRIID) have the unique resources to handle highly infectious agents and the ability to identify specific agent strains (CDC, 2005). Reference laboratories, sometimes referred to as confirmatory reference, can perform tests to detect and confirm the presence of a threat agent. These laboratories ensure a timely local response, rather than having to rely on confirmation from CDC labs. Sentinel laboratories represent the thousands of hospital-based and some veterinary labs that are in the front lines. Their responsibility is to refer a suspicious sample to the right reference laboratory. For instance, in the case of severe acute respiratory syndrome (SARS), CDC laboratories identified the unique DNA sequence of the virus that causes the disease. LRN then developed tests and materials needed to support these tests and gave LRN members access to the tests and materials (CDC, 2005).

Veterinary Diagnostic Laboratories

Each state has a publicly funded veterinary diagnostic laboratory. The sizes and diagnostic capabilities of these laboratories vary widely, ranging from a few states with rudimentary laboratories that perform only serology for one or more eradication program diseases such as brucellosis to state systems that offer a complete range of diagnostic services for all economically important agricultural animal species, companion animals, and wildlife in multiple locations within the state. Some laboratories specialize by species, primarily serving the local needs within a specified geographic area, but the majority of state laboratories cover a broad range of species and conditions. Many are directly affiliated and co-located with a university-based college of veterinary medicine or veterinary science department.

The state laboratory system is represented nationally by the AAVLD. One of AAVLD's cardinal activities is to accredit publicly funded veterinary diagnostic laboratories. The accreditation program meets international standards established by the World Animal Health Organization (through the OIE). While the OIE does not conduct an accreditation program, the Standards Commission does provide standards to member countries as a guide for accrediting laboratories conducting assays for infectious diseases on the OIE lists ("OIE Quality Standard and Guidelines for Veterinary Laboratories: Infectious Diseases"). The OIE reference standards, with minor exceptions, reflect ISO17025 laboratory accreditation standards.

There currently are 38 AAVLD accredited, full-service laboratories/ systems in the United States, located in 34 states, and two AAVLD accredited laboratories in two Canadian provinces. Accredited laboratories undergo a site visit every 5 years unless major changes in funding or personnel warrant an earlier visit. Importantly, accredited laboratories maintain a full cadre of disciplinary specialists and laboratory sections that include pathology, bacteriology, virology, parasitology, and toxicology. In some cases, a few of these services are contracted to another accredited laboratory. Nearly all accredited laboratories have board-certified specialists who head laboratory sections. Specialized molecular assays such as PCR are in common use, and complete workup of unusual and challenging diagnostic cases is routine. The broad capability of these laboratories was an underappreciated resource for both diagnostic testing capacity and capabilities. One example is the wealth of expertise and equipment in toxicology. As the nation's public health laboratories struggle to prepare for the potential challenge of toxins, whether biological or chemical, intentionally introduced in environmental or food matrices, the board-certified veterinary toxicologists, analytical chemists, sophisticated equipment, and familiarity with many of these potential toxins extant in accredited veterinary diagnostic laboratories simply must not be overlooked.

Although state laboratories offer a rich resource of diagnostic services and data, and the accreditation process ensures the quality of these services and data, a lack of uniformity among laboratories limits the value of compiled data. This is primarily due to the use of different standard operating procedures and assays that, although withstanding the test of time, have not undergone rigorous validation to meet current international standards. To that end, the AAVLD recently began a process of compiling a table of approved methods to use as a standard for accreditation. Once completed, standard operating procedures for each approved method will be available to all, helping to promote standardization nationwide and to increase the value of diagnostic data routinely generated in state laboratories.

The size and quality of capital assets in the state laboratories vary widely, from aging buildings in which it is difficult to meet current standards for security and biocontainment, to recently completed new buildings incorporating the latest standards for good laboratory practice, security, biocontainment, and waste disposal. Capital infrastructure in many of the state laboratories is in need of modernization, but most state budgets currently struggle to maintain the current buildings and cannot begin to address new capital investments required to meet contemporary standards. In an unpublished 2003 informal survey conducted by the AAVLD to assess Biosafety Level 3 (BSL-3) laboratory and necropsy capacity in state laboratories, 23 laboratories of 35 responding (from 33 states and most accredited laboratories) had a BSL-3 laboratory, ranging from 200 to 2,000 net square feet (Terry McElwain, AAVLD, unpublished data, 2003). However, there were no BSL-3 necropsy suites in any of the state laboratories that responded. Some had remote access to a BSL-3 necropsy laboratory in another location, primarily associated with Animal Biosafety Level 3 (ABSL-3) animal housing. In 2004, one state laboratory completed a new facility that has a large BSL-3 necropsy suite, and at least one other new state laboratory, to be completed in 2006, will also have a BSL-3 necropsy suite. For additional information on BSL-3 laboratories, see the section on Containment Facilities later in this chapter.

Commercial Laboratories

Over the past 10 years, commercial for-profit laboratories have moved forcefully into the realm of animal diagnostics. Initially the primary focus of these laboratories was on companion animal pathology, both clinical and anatomical. More recently, commercial laboratories have captured an increasing share of the routine serological and microbiological work that in the past was performed almost entirely by federal and state laboratories. The availability of approved assay kits for diagnostic work in animals and the USDA policy of approving small laboratories in veterinary practices and other private labs to perform testing for reportable diseases such as equine infectious anemia has facilitated this change. Few of these laboratories have the capability to perform routine necropsies and few offer viral culture or toxicology services. This has placed state laboratories in an interesting paradox, because necropsy, toxicology, and virology have not been profitable services for state laboratories. Thus, like their commercial counterparts, state laboratories have relied on high-volume services such as serology to enhance their revenues. In addition, commercial laboratories may not always have operational relationships with regulatory agencies as seamless as those of state laboratories. As a consequence, test results in these laboratories may not be as readily available for analysis in passive surveillance programs (except for results on reportable or notifiable diseases).

Development of private laboratories in corporate food animal production systems has also impacted animal disease diagnostics. This trend is due, at least in part, to the technological advances that have made commercial kits widely available, but it is also driven by the development of hazard analysis and critical control point (HACCP) and other quality assurance programs and by the confidentiality necessary in the competitive world of food production. The development of in-house laboratories has been most notable in the swine and poultry industries, but to some extent it is also practiced in commercial aquaculture and other industries. Again, laboratory results are not available publicly, and the development of in-house diagnostics in some production systems has replaced consultations with health professionals. For example, a 2000 study revealed that over 20 percent of swine production units (primarily small operations) reported no veterinary visits in the previous 12 months (USDA APHIS-VS, 2001). Rapid recognition, diagnosis, and reporting of health problems arising from the introduction of exotic agents are absolutely essential for limiting the spread of infectious diseases. The development of vertical integration in laboratory analysis thus provides a special challenge in disease control. The extent and effectiveness of quality assurance programs in private laboratories is difficult to assess. Opportunities for outside review and oversight of these laboratories are limited unless they utilize the International Organization for Standardization (ISO) accreditation or some other system of assessment by auditors, since AAVLD offers accreditation only for publicly funded laboratories. The USDA does assess laboratory practices before approving laboratories to run assays for reportable diseases.

ANIMAL HEALTH RESEARCH

Research on animal health issues is funded by a variety of means and at a range of levels, usually depending on three main factors: the affected species, the degree of relevance to human health, and the economic impact of the animal disease. At one end of the spectrum, disease issues impacting only companion animals or wildlife and with no relevance to human health are traditionally very poorly supported. Examples here might include coronavirus infection in cats (prior to the outbreak of SARS) or parasitic infections of wildlife. In the absence of federal support, private charities and foundations, academic institutions, and industry typically fund research on these issues, and usually at a very modest level (Eisner, 1991). Companion animal research is conducted primarily by pharmaceutical and food companies that have research and development units; this research is usually product-oriented research that covers the span from basic research all the way to clinical trials.

At the other extreme, animal health issues that have direct relevance to human health, as well as significance to animal populations, are usually funded by federal human health sources (National Institutes of Health), and often at munificent levels. Examples are bovine spongiform encephalopathy and highly pathogenic avian influenza, two diseases that impact very economically important agricultural species, affect international trade, but also spell possibly fatal outcomes in humans. Falling in between these two ends of the spectrum are the majority of animal health problems, with USDA, and more recently DHS, supporting most of the research.

In 1994, the Federal Crop Insurance Reform and Department of Agriculture Reorganization Act authorized the Secretary of Agriculture to appoint an undersecretary for research, education, and economics (REE). Four agencies were brought into the REE mission, including the Agricultural Research Service (ARS), the CSREES, the Economic Research Service (ERS), and the National Agricultural Statistics Service (NASS). The REE agencies are also required to work with the USDA action and regulatory agencies in support of their operations and missions (U.S. Congress, 1994).

The most recent REE Strategic Plan identifies five key outcomes as goals of its research effort: (1) a highly competitive global agricultural system; (2) a safe and secure food and fiber system; (3) healthy, well-nourished children, youth, and families; (4) greater harmony between agriculture and the environment; and (5) enhanced economic opportunity and quality of life for citizens and communities (USDA, 2002d).

In 2002, ARS, CSREES, and ERS allocated \$120 million, \$39 million, and \$0.65 million, respectively, for animal health research (Karen Lawson, personal communication, 2005). In 2001, REE agencies collectively employed 4,132 science-related technical staff, with the largest portion (approximately 75 percent) employed by ARS (USDA, 2001). In 2004, ARS employed 282 scientists in its animal health and protection programs (USDA ARS, 2004).

As the largest REE agency in overall program and staff budget (approximately \$1 billion) and as the principal in-house research agency for the USDA, the ARS uses its funds to support a blend of basic and applied research activities (NRC, 2003a). Through its 22 national programs, the ARS has created a substantial infrastructure of research laboratories across the United States with 244 ARS laboratories at 103 locations and 41 work sites. The laboratories, over 100 of which are used for research on agricultural animals, include over 3,000 buildings and the agency covers 400,478 acres of land dedicated to research (GAO, 2000).

The major source of funding for university-based research is CSREES, which in 2002 invested approximately $29 million in animal health research at universities out of $1.04 billion appropriated for all research and other activities (Peter J. Johnson, personal communication, 2005). In 2001, CSREES employed 162 science-related technical staff to oversee the funding of these extramural research activities. The research and education activities of CSREES were originally authorized under the Hatch Act of 1887. Under its current authorities, CSREES assists research and education programs at state institutions, including state agricultural experiment stations, schools of forestry, 1890 colleges, land-grant institutions, colleges of veterinary medicine, and other eligible institutions.

CSREES is also charged with implementing USDA's higher education mission in the food and agricultural sciences. The cooperative extension system is a national educational network of partners from CSREES, land-grant university cooperative extension services, and cooperative extension services in the 3,150 counties of the United States. The work of the cooperative extension service was originally authorized by the Smith-Lever Act of 1914 (U.S. Congress, 1914). The educational arm of CSREES represents an important function in transferring knowledge produced by researchers to agriculturalists who could then apply research results to improve production and resolve problems. CSREES has identified 59 programs that span the biological, physical, and social sciences and that are related to agricultural research, economic analysis, statistics, extension, and higher education.

In 2004, CSREES received funding of $1.124 billion. Just under half of the research funds administered by CSREES are "formula funds," based on formulas related to the size of rural and farm populations and distributed to state agricultural experiment stations. Under formula funds, the Animal Health Research and Disease Program (Section 1433) received only $4.5 million in 2004. Competitive, peer-reviewed grants for research and education programs make up approximately 25 percent of CSREES research funds. The National Research Initiative (NRI) grants are part of this external funding and in 2004 were appropriated at $164 million. The NRI is a competitive grant and peer-review research program. Some of these funds were used for research to protect natural resources including wildlife, optimize livestock health and productivity, and protect human health and food safety. Special grants are also used for selected projects and are largely based on congressional directives.

Since the establishment of DHS, additional funding has become available for animal health issues that are considered of national security interest. This includes most foreign animal diseases, as a deliberate introduction of one of these diseases could have severe economic consequences. Recently, two different DHS Centers of Excellence, a Center for Foreign Animal and Zoonotic Diseases, and a Center for Food Safety, were established at land-grant universities.

A forthcoming NRC report *Critical Needs for Research in Veterinary Science* (NRC, 2005) will examine the broad range of fields in which veterinary researchers can contribute, including research in comparative medicine. Diseases cause a significant amount of morbidity and mortality in both people and animals. To help alleviate this suffering, biomedical research has included the use of animals as one component of research to understand, treat, and cure many human and animal diseases. The parallels between animal physiology, genetics, and pathology have been noted for a long time, and the study of animals to understand human disease is also an accepted area of biomedical research. Animals develop many of the same diseases as humans and are susceptible to many of the same bacteria, viruses, and parasites. Animal models have been used successfully for research when they share similar and relevant characteristics with people. Comparative medical research uses animals to simulate biological functions and help link molecular, cellular, tissue, organ systems, and organism systems (NRC, 2004d). Unlike discipline-oriented researchers, comparative medical scientists bridge the interface between fundamental and basic science and human health. Beyond the benefits to both of these research areas are the direct benefits to animals themselves.

However, the present involvement of veterinarians in comparative medicine is insufficient, and current trends to support that participation are disconcerting. Federal funds for animal research have been relatively static and the prospects for significant increases in the future seem unlikely. While the National Institutes of Health (NIH) and National Science Foundation (NSF) have enjoyed substantial increases in funding, animal health has largely benefited as a by-product from the study of animal models and comparative medical systems. Concern persists about the lack of new animal scientists and researchers for both animal and biomedical research in the future. The contemporary problems of food safety, zoonotic diseases, emerging diseases, and agro- and bioterrorism have stimulated interest in these fields of research, but accelerated interdisciplinary and new intergovernmental programs have been slow to develop.

If the contemporary problems brought about by the convergence of human and animal health are to be adequately addressed, comparative studies will need a national focus and sustained attention.

Scientific research and the investigators who carry out that work must be a central part of the animal health framework. Such research is critical to reducing animal disease and suffering and to the development of new products, treatments, and techniques for animals that indirectly benefit society. The efficiency and productivity of animal agriculture over the years has been a function of successful research on animal nutrition, effective production systems, and reducing the incidence of animal diseases.

Containment Facilities

Of particular note in the context of discussing research on animal health is the issue of containment facilities. Studies of infectious diseases, whether of interest solely for animal health or as animal models of human disease, need to be undertaken in a manner that ensures safety for the operator as well as the general public. As such, there are specified containment levels for the various organisms that mandate certain structural and procedural necessities. Containment facilities are classified as Biosafety Levels 1 through 4, with 4 being the most restrictive (HHS, 1999). Biosafety level 3 (BSL-3 or BSL-3 Ag) provides the high degree of containment that is needed when studying a variety of organisms with a recognized potential for significant detrimental impact on animal or human health or on natural ecosystems (Box 2-3).

This level of containment requires stringent measures such as protective clothing and respirators; filtered air supply and exhaust; sterilization of materials originating from the facility, including animal waste; and strictly controlled entry and exit. The number of BSL-3 laboratories in the United States is limited; in particular, there are very few BSL-3 Ag entities due to their demanding and expensive engineering and construction requirements (USDA, 2002b). Consequently, even with full institutional volition and funding to undertake research with certain agents, such as classical swine fever, monkeypox, or tularemia, studies can only be conducted if the building meets the design standards required. In 2003, the National Institute of Allergy and Infectious Disease announced awards to 11 universities to build biosafety laboratories as part of a regional and national network for infectious disease research under its Biodefense Research Agenda. Of those selected, two of the regional centers are associated with veterinary science: the University of Missouri-Columbia and Colorado State University. Each plans to build a BSL-3 facility.

BOX 2-3
Definitions of Level 3 Biocontainment Facilities in the Animal Health Framework

Animal Biosafety Level 3 (ABSL-3): Involves practices suitable for work with animals infected with indigenous or exotic BSL-3 agents that present the potential of aerosol transmission and of causing serious or potentially lethal disease. ABSL-3 builds upon the standard practices, procedures, containment equipment, and facility requirements of ABSL-2.

Biosafety Level 3 (BSL-3): Used with agents that may be indigenous or exotic to the United States that can be contracted by the respiratory route and may cause serious or lethal diseases to humans or animals or cause moderate economic loss to the animal industries. The BSL-3 facility is designed to support research activities with serious or potentially lethal biohazardous materials or infectious substances.

Biosafety Level 3 Agriculture (BSL-3Ag): Designation for animal facilities in which research involves BSL-3 biological agents that present a risk of causing great economic harm should they infect the indigenous animal population (e.g., foot-and-mouth disease). Using the containment features of the standard BSL-3 facility as a starting point, BSL-3Ag facilities are specifically designed to protect the environment by including almost all of the features ordinarily used for BSL-4 facilities as enhancements. All BSL-3Ag containment spaces must be designed, constructed, and certified as primary containment barriers. Colloquially, they may be referred to as ABSL-3 Ag.

SOURCE: Biosafety in Microbiological and Biomedical Laboratories, 4th edition. Available at http://www.cdc.gov/od/ohs/biosfty/bmbl4/bmbl4toc.htm

INTERNATIONAL ISSUES

Deterrence and prevention of animal disease in the United States involve global strategies that are directed at reducing a potential threat before it reaches the U.S. borders and a border strategy that focuses on interdicting a threat agent at U.S. ports of entry (NRC, 2003a). An overview of international organizations involved in prevention, detection, and diagnosis is provided earlier in this chapter, so the discussion below focuses on components of the framework responsible for interdicting threat agents at U.S. ports as well as during the sale and transport of animals (particularly exotic animals) once they have entered the country.

Importation, Sale, and Transport of Animals

In 2003, the United States exported 125,000 head of cattle and imported about 1.52 million head; there were 134,000 live hogs exported and 7.25 million live hogs imported (Beghin et al., 2004). Every year, a variety of sources provides millions of animals to the exotic companion animal trade. Animals are captured from their native habitat and transported to various countries to be sold as companion animals. Others are surplus animals from zoos or their offspring. Backyard breeders also supply exotic companion animals (API, 2003). Consequently, the importation of animals is an important concern of the animal health framework.

In 2002, more than 22 federal agencies were consolidated into the Department of Homeland Security (DHS), including components of APHIS that conduct inspection and animal quarantine activities at U.S. ports and the Plum Island Animal Disease Center (PIADC). Approximately 2,600 employees from the APHIS Agriculture Quarantine and Inspection (AQI) force became part of the DHS Border and Transportation Security's Bureau of Customs and Border Protection (CBP) on March 1, 2003 (USDA APHIS, 2003a).

Although DHS is now responsible for protecting the nation's borders, USDA APHIS, continues to set agricultural policy through risk assessment, pathway analysis, and rule making, including specific quarantine, testing, and other conditions under which animals, animal products, and veterinary biologics can be imported. These policies are then implemented by DHS (USDA APHIS, 2003a). USDA APHIS-VS port veterinarians inspect live animals at border ports and place animals in quarantine until testing is completed. They are located at 43 VS areas and report to the veterinarian in charge of the VS-Area Office (Joseph Annelli, personal communication, April 2004). With agricultural border inspectors now a part of DHS, VS has identified a need for developing new protocols for training and interacting with these inspectors, as well as a need to work with DHS to implement improvements recommended in the Animal Health Safeguarding Review regarding pest exclusion activities at U.S. borders in its strategic plan (USDA, 2004d).

The Secretary of the Department of Health and Human Services (HHS), through the CDC, has the authority to make and enforce regulations to prevent transmission of infectious disease from foreign countries into the United States (42 CFR70 and 71). Under these regulatory authorities, CDC has established embargoes on prairie dogs and other animals that could carry the monkeypox virus and on birds from specified Southeast Asian countries (CDC, 2003d; CDC, 2004b). Table 2-2 provides a summary of agencies and functions involved in border control and a review of the events related to their organization.

TABLE 2-2 DHS Border and Transportation Security (BTS), Bureau of Custom and Border Protection (CBP), and Other Components Addressing Animal Diseases

Agency	Agency Description, Responsibilities, & Major Events
Border and Transportation Security (BTS)	• The largest of the 5 DHS directorates. • Includes former U.S. Customs Service, border security function/enforcement division of INS, APHIS, Federal Law Enforcement Training Center, and the Transportation Security Administration. • Responsible for securing the nation's air, land, and sea borders. • Responsible for securing the nation's transportation systems and enforcing the nation's immigration laws.
Bureau of Custom and Border Protection (CBP)	• March 1, 2003, approximately 42,000 employees were transferred from U.S. Customs Service, INS, and APHIS to the new CBP, a new agency under the BTS directorate within the DHS. • Approximately 2,700 former USDA employees from the AQI program and APHIS were transferred into DHS. • Former APHIS-PPQ personnel at ports of entry (POEs) who were directly involved in terminal/plane inspections (100% time) were transferred to DHS; those with 60-70% time not doing inspection at terminals/planes were not transferred. • The agricultural import and entry inspection functions that were transferred include: reviewing passenger declarations and cargo manifests to target high-risk agricultural passengers or cargo shipments. • The new CBP also carries out the traditional missions of the predecessor agencies making up CBP (seizing illegal drugs and other contraband at the U.S. border; apprehending people who attempt to enter the U.S. illegally; detecting counterfeit entry documents; determining the admissibility of people and goods; protecting U.S. agricultural interests from harmful pests or diseases; regulating and facilitating international trade; collecting duties and fees; enforcing all laws of the United States at borders).
Office Field Operations (OFO)	• Oversees over 25,000 employees at 20 field operation offices (OFOs), 317 POEs, and 14 preclearance stations in Canada and the Caribbean. • Responsible for enforcing customs, immigration, and agriculture laws and regulations at U.S. borders. • Manages core custom and border protection programs

continued

TABLE 2-2 Continued

Agency	Agency Description, Responsibilities, & Major Events
Border and Transportation	• The largest of the 5 DHS directorates. (i.e., border security and facilitation, interdiction and security, passenger operations, targeting analysis and canine enforcement; trade compliance and facilitation, trade risk management, enforcement, and seizures and penalties as well as examine trade operations to focus on antiterrorism). • Administer Agricultural Inspection Policy and Programs (agricultural quarantine inspection, AQI, at all ports of entry in order to protect the health of U.S. plant and animal resources). • Administer immigrations policy programs. • Annual operating budget of $1.1 billion. • Each OFO is run by a Director of Field Operations (DFO)
OFO - Associate Commissionerof Agricultural Inspection Policy and Programs	• Policy advisor to the Office of the Commissioner on all agricultural issues.
CBP Port Director	• On March 1, 2003, CBP designated one port director at each port of entry in charge of all federal inspection services establishing a single, unified chain of command.
CBP Ag. Specialist	• Enforce USDA regulations and seize any articles in violation of regulations • Conduct prearrival risk analysis. • Cargo examination for quarantine disease and pests. • Collection, preparation, and submission of pest and disease samples to USDA. • Seizures, safeguarding, destruction, or reexportation of inadmissible cargo. • Negotiation of compliance agreements with importers of regulated commodities. • Stationed only at ports of entry with large volumes of cargo and only to support the CBP officers. • As of October 4, 2003, there are 1,471 full-time permanent agricultural inspectors on board. • New CBP officers will be trained at the Federal Law Enforcement Training Center (FLETC) in Glynco, Ga., and agricultural specialists will continue to learn their trade at PPQ Professional Development Center in Frederick, Md. • Agricultural training of CBP officers highlighted as a concern.
CBP and FDA	• In October 2003, CBP and FDA entered into an agreement to further protect U.S. food supply.

	• At ports of entry, CBP inspectors now carry out special inspection and sampling of foreign food imports and make referrals back to FDA for further testing and analysis. • CBP and FDA work side by side in targeting efforts, making joint decisions about any food shipments that could pose a potential threat to the United States.
National Targeting Center (NTC)	• Part of CBP's OFO, the NTC provides tactical targeting and analytical research support for antiterrorism efforts to DHS and its Operations Center. NTC has representatives from all CBP disciplines.
CBP Laboratories and Scientific Sciences Division (LLS)	• On December 8, 2003, LLS moved its Radiation Portal Monitor to the NTC.

SOURCES: DHS, March 2004b; Bonner, 2004; USAHA, 2003a; CBP Today, March 2003; US CBP website press release, January 2004; Khawaja Ahmad, USDA-APHIS-VS, personal communication, April 2004; DHS 2004e.

ADDRESSING FUTURE ANIMAL DISEASE RISKS

A critical tool for informing decisions about how to prevent or respond to animal disease is the evaluation of risk related to the potential occurrence, transmission, or establishment of animal diseases. In the context of animal disease, risk analysis is the framework for understanding the impact of a wide variety of variables on animal health, and particularly, of the transmission of disease through the movement of animals, animal products, and vectors.

With increased globalization and increasing access to foreign animal markets, the avenue of contamination through importation of animals and products that harbor infectious agents requires constant attention. The WTO/SPS Agreement described earlier in this chapter emphasizes the use of scientific principles as a basis for the implementation of animal and human health-related protection measures in trade. Signatory nations must document the risk that is posed by importing another country's products in order to justify trade barriers or sanitation requirements erected to safeguard domestic animals. The agreement employs the term "risk assessment" (one component of the risk analysis process) as "the evaluation of the likelihood of entry, establishment or spread of a pest or disease within the territory of an importing member according to the sanitary and phytosanitary measures which might be applied, and of the associated potential biological and economic consequences; or the evaluation of the potential for adverse effects on human or animal health arising from the presence of additives, contaminants, toxins, or disease-causing organisms in food, feedstuffs and beverages."

According to the OIE (2003), identifying the pathogenic agents associated with the importation of a commodity that could potentially produce adverse consequences is the first component of risk analysis. The second component is risk assessment, described as a process of four interrelated steps:

> Step 1—Release Assessment: Description of the biological pathway(s) necessary for an importation activity to "release" pathogenic agents into a particular environment, and estimating the probability of that process occurring, either qualitatively and/or quantitatively.
>
> Step 2—Exposure Assessment: Qualitative and/or quantitative description of the biological pathway(s) necessary for exposure of animals and humans in the importing country to the hazards released from a given source and estimating the probability of exposure.
>
> Step 3—Consequence Assessment: Description of the potential consequences of a given exposure and estimates the probability of them occurring. Examples of direct consequences include animal infection, disease, and production losses, and examples of an indirect consequence would be potential trade losses or compensation losses.
>
> Step 4—Risk Characterization or Risk Estimation: Integration of all of the information gathered during the risk assessment process is integrated to produce overall measures of risks associated with the hazards identified at the outset. An example of a final output might be estimated numbers of herds, flocks, animals, or people likely to experience health impacts over time.

The OIE framework is useful for considering and integrating the complexities of risk assessment into logical steps that can be better analyzed both qualitatively and quantitatively. Two additional components of risk analysis are (1) risk management, the process by which the results of risk assessment are integrated with other information, such as political, social, economic, and engineering considerations, for example, to arrive at decisions about the need and methods for risk reduction; and (2) risk communication, the explanation of findings from the risk assessment to risk managers, consumers, industry, and other interested parties in an interactive dialogue about risk-related factors and perceptions (NRC, 1994b). While other tools and frameworks continue to emerge, this systematic approach is important for making decisions, setting priorities, planning interventions, and evaluating prevention and control strategies.

The nation's animal and public health could benefit more fully from the potentially powerful analytic capabilities of risk analysis and risk assessment if there were more widespread understanding of and participation in the way in which mathematical models of disease incidence and spread are used to determine risk, and the assumptions inherent in those models. By formalizing and strengthening links and communication between risk assessment modelers, who understand risk modeling methodologies, and biologists, who understand the agents, animals, biology, and pathways of the disease, the biological accuracy of pathways being modeled could be improved. These interactions might help, for example, to clarify uncertainty in the prevailing knowledge of the disease and in the pathways being modeled, which are never known with absolute certainty, in order to understand the confidence that should be applied to reported risk estimates.

In general, there is a need to promote the education of risk assessment methods for decision-makers, biologists, diagnosticians, and others who will be called upon to use, or to respond to, risk assessment reports. New ways to communicate key findings and conclusions of each of the four steps of risk assessment to those who are neither risk modelers nor experts in the disease are needed, so that those who must apply the results of risk assessments to policy or action can better interpret the bases for risk assessment results and can have confidence in their understanding of the strengths and weaknesses of the methods used.

That goal might be accomplished, in part, by incorporating the specific goals and objectives of decision-makers and animal health planners into initial stages of risk assessment design so that risk assessments can be more focused and directed, and thus more precise, in addressing specific animal health issues.

EDUCATION AND TRAINING

Education and Training of Veterinarians

Veterinary Schools

Veterinary medicine comprises several distinct fields of practice, including the care of various species of food-animals, small animals, equids, general or rural practice (mixed domestic animals), ecosystem health (including wildlife disease and conservation biology), public health, and biomedical science. Not surprisingly, veterinary schools face a difficult challenge in producing sufficient graduates for all of these fields with the appropriate depth of competence across the full range of veterinary practices.

The United States has 28 schools of veterinary medicine that graduate approximately 2,000 individuals each year with a Doctor of Veterinary

Medicine (DVM) degree (AAVMC, 2003). Attaining this degree requires a minimum of 2–4 years of university preparation followed by a professional curriculum that normally extends over 4 years. This pattern of education emulates human medical education, with a key difference being that internship is not required prior to licensing for veterinarians. The system has served veterinary medicine reasonably well in the past, but it has not changed in about 50 years despite recent enormous changes in society that have generated markedly altered production systems and disease patterns.

The students entering veterinary schools and their decisions to specialize are also changing. For example, veterinary students are increasingly from urban environments and are women. Another trend is that, with more disposable income and greater expectations for the level of care and services for their animals, companion animal owners have demanded greater sophistication and improved health care delivery that has resulted in specialization into services such as oncology, critical care, internal medicine, and ophthalmology. These dramatic increases in specialization in companion animal services and practices, and improved financial rewards, have influenced student decision making to enter these fields.

DVM programs are uniformly subjected to accreditation by the American Veterinary Medical Association (AVMA), which sets "standard requirements of an accredited or approved college of veterinary medicine" (AVMA, 2004a). These standards include those relating to organization, finances, physical facilities and equipment, clinical resources, library and information resources, students, admission, faculty, curriculum, research programs, and outcomes assessments. The AVMA's Council on Education reviews each veterinary school every 7 years.

The Association of American Veterinary Medical Colleges (AAVMC) provides a collective voice for the veterinary schools (AAVMC, 2004). It publishes the *Journal of Veterinary Medical Education*, sponsors biennial symposia, manages a national veterinary student application process, and provides leadership in addressing current issues in veterinary education and research.

Licensing

State agencies license veterinarians. U.S graduates must be from an AVMA accredited school and have passed a standard North American Veterinary Licensing Examination (NAVLE) to enter private practice. While there is some opportunity for students to focus their undergraduate clinical training in one of the specific fields of veterinary medicine, accreditation requirements and the broad range of subject matter covered by the NAVLE puts a limitation on the practical extent of such training. As a consequence, specialization in the various fields of veterinary medicine occurs at the postgraduate level. The relatively modest incomes that are the norm in veterinary medicine (with means that range from $84,000 to $92,000/year for different fields in 2002), together with high levels of student indebtedness (a mean of $71,000 in 2002) may deter new graduates from opting for postgraduate training. This has led to the suggestion that veterinary educators should consider an engineering model of undergraduate professional education in which veterinary students elect a curriculum track with the depth of study in different disciplines appropriate to the field of their choice (Eyre, 2002; Radostits, 2003; Nielsen, 2003). This would require a change in licensing policies, which has been advocated by some (Karg, 2000).

Training in Population Health/Food Systems

An adequate education in population health is essential for veterinarians on the first line of defense in dealing with animal diseases in the livestock industry as private practitioners and as employees of government agencies or commercial enterprises. It is also essential for those involved in the food system, public health, and ecosystem health. However, since the objective of about 75 percent of students is to enter companion animal practice or a related specialty, present curricula emphasize individual animal medicine.

A symposium of U.S. veterinary educators (Hird et al., 2002) held in 2002 concluded that:

> A crisis in veterinary medicine exists that requires urgent action from veterinary educators, veterinary associations and organizations, and public and private practitioners. The convergence of animal, human, and environmental health issues has created the need for veterinarians with a level of knowledge and skills that is not being achieved by either new graduates or the current pool of veterinarians. Unprecedented changes in food animal production and health, human and animal demographics, diseases, concern for animal well-being and welfare, antibiotic resistance, and biotechnology are occurring. In addition increasing threats to animal populations from the introduction of exotic animal diseases, either accidentally or intentionally, require a much larger cadre of veterinarians with training in population health concepts if the US is to manage exotic disease outbreaks and maintain the security of the of the US food supply.

The conclusions that emerged from this symposium echoed similar ones made in the 1972 NRC report *New Horizons in Veterinary Medicine* and the 1988 Pew *National Veterinary Education Program* (NRC, 1972; Pritchard, 1988).

Training in Public Health

A recent survey of education in public health in 27 (of the 28) U.S. veterinary schools found that the curricula of all 27 required at least one course in public health; when epidemiology was included, the contact hours assigned to these subjects ranged from 30 to 120 (mean of 67). Only four schools, however, have required clinical rotations in public health (Riddle et al., 2004). Twenty-four of the 27 schools offer from one to six elective courses varying from a total of 15 to 288 hours. Eight schools offer elective clinical rotations of 3–4 weeks in length. Twenty-three schools offer some form of advanced training in public health or epidemiology, four offering a dual DVM/Masters of Public Health program. Fifteen schools offer or are about to offer some form of DVM program combined with an advanced degree related to public health. Statistics describing first-year employment of new graduates (Table 2-3) indicate that few, if any, opt for careers in public health or have the opportunity without further education. Leaders in veterinary education have called for the profession and its educational establishment to give much more attention to meeting societal needs in this field (Hoblet et al., 2003).

TABLE 2-3 First-Year Employment, 2004 Veterinary Graduates in Various Fields

	Percent
Private Clinical Practice	68.2
Large animal exclusive	2.5
Large animal predominant	2.8
Mixed animal	9.0
Small animal exclusive	40.4
Small animal predominant	10.0
Equid	3.4
Public or Corporate Employment	1.9
College/university	0.1
Uniformed services	1.2
Federal government	0.1
State/local government	0.1
Industry/commercial	0.1
Not-for-profit	0.2
Other	0.6
Unknown	3.7
Advanced study programs	25.7

SOURCE: AVMA, 2004b

Ecosystem Health

Ecosystem health provides a broad context for veterinary education to address wildlife diseases and conservation biology at the level of multiple populations that share the same environment (Van Leeuwen et al., 1998; Deem, 2004). While some schools offer undergraduates the opportunity to choose elective courses or rotation in wildlife diseases or zoological medicine, most new graduates who wish to specialize in wildlife diseases undertake postgraduate studies to this end. Veterinary schools in Canada have jointly created an innovative elective undergraduate rotation in ecosystem health (Ribble et al., 1997).

Veterinary Technology Programs

Veterinary technicians are important members of veterinary practice teams, government agencies, biomedical research laboratories, diagnostic laboratories, and commercial enterprises. Opportunities to make rural practice more attractive could depend on having veterinary technicians who are better suited and empowered to provide appropriate support to veterinary practitioners. There are 104 programs in veterinary technology in the United States accredited by the AVMA; 15 offer baccalaureate degrees, 2 of which are at a veterinary college.

Postgraduate Studies

Although data on the total number of graduate students in the veterinary sciences are unknown, a 2004 AVMA survey indicated that of 2,225 College of Veterinary Medicine (CVM) graduates, roughly 25 percent responded they were entering graduate studies at CVMs and elsewhere (Shepherd, 2004). In 2002, 27.7 percent of all female graduates and 23 percent of all male graduates directly entered advanced studies, including internships, residencies, and graduate training programs (Wise and Shepherd, 2004).

In order to encourage more veterinary students to opt for postgraduate training, at least 10 veterinary schools offer combined DVM/graduate degree programs, such as DVM/PhD and DVM/Master's programs, not counting schools with joint MPH programs (Riddle et al., 2004). Several colleges have recently offered new DVM/MPH dual degree programs that can be completed in 4 years.

Postgraduate training can be of several types: (1) clinical training, leading to certification as a specialist (or diplomate); (2) research training, to prepare the veterinarian to be an independent research scientist in a specific area, such as immunology, physiology, epidemiology, microbiology and toxicology; this training may or may not lead to a PhD, although individuals seriously interested in a research career typically pursue a PhD, followed by postdoctoral training; and (3) a combination of research and clinical training—for example, veterinary pathology or laboratory animal medicine.

Clinical Training

Many, if not most, new veterinary graduates seeking formal postgraduate education elect residency training (which may be in conjunction with a M.Sc. degree) and board certification in a medical discipline with a view to becoming a clinical specialist, often in the companion animal health discipline.

In the United States, the AVMA guides and regulates the formal processes for clinical specialization in a veterinary medical discipline, a process that began in 1949 with the pathology specialty. There are now 20 specialty colleges (See Table 2-4). Veterinarians who wish to achieve the status of a specialist in a medical discipline must undertake an approved residency program and subsequently pass a rigorous examination set by a recognized specialty college. Those who successfully complete a program become registered "diplomates" in the college or board they choose. It normally requires about 3–5 years for a new graduate to achieve this goal. Specialization by species was resisted for many years, except in the case of laboratory animals, where the American College of Laboratory Animal Medicine has existed since 1957. The American College of Poultry Veterinarians was established in 1991. The American Board of Veterinary Practitioners (established in 1976) recently provided categories for specialization in avian practice, beef cattle practice, dairy practice, and swine health management. The number of diplomates in each of these categories is modest, ranging from 11 to 107.

Diplomate status in a specialty college has become a required qualification for faculty in clinical departments of many of the nation's faculties of veterinary medicine and has greatly enhanced the quality of clinical education. The diplomate status in one of several disciplines is the preferred qualification for section heads in diagnostic laboratories that opt for accreditation by the AAVLD. Increasing the strength of the nation's animal diagnostic laboratory and field investigative network will depend in part on having adequate numbers of veterinarians with specialist qualifications in pathology, epidemiology, microbiology, toxicology, and wildlife diseases, as well as other laboratory professionals.

TABLE 2-4 Active, Board-Certified Diplomates (as of December 2004)

Field	No. of Diplomates	
All Fields	7,970	
Anesthesiologists	148	
Animal Behaviorists	36	
Dentistry	75	
Dermatologists	158	
Emergency and Critical Care	156	
Internal Medicine	1,478	
Cardiology		120
Internal Medicine, Small Animal		788
Internal Medicine, Large Animal		357
Neurology		126
Oncology		151
Laboratory Animal Medicine	677	
Microbiologists	164	
Bacteriology/Mycology		33
Immunology		43
Microbiology		85
Virology		50
Nutrition	47	
Ophthalmologists	264	
Pathologists	1,411	
Anatomical Pathology		1,210
Clinical Pathology		255
Toxicological Pathology		38
Pharmacology	43	
Poultry	247	
Practitioners	740	
Avian		107
Beef Cattle		11
Canine and Feline		408
Dairy		30
Equine		74
Feline Exclusive		71
Food Animal		20
Swine Health Management		18
Preventive Medicine	531	
Epidemiology		64
Radiology	264	
Radiation Oncology		34
Veterinary Surgeons	1,041	
Small Animal		43
Large Animal		23
Theriogenologists	306	
Toxicology	98	
Zoological Medicine	83	

SOURCE: AVMA, 2004c.

Research Training and Combination Training

Currently, students who seek board certification are encouraged to pursue a PhD if they have an interest in research. Unlike MD equivalents, who often enter postdoctoral training in a research environment, opportunities for rigorous DVM postdoctoral research training are few. While it is not unusual for an MD involved in research not to hold a PhD, it is still expected by veterinary colleges that a veterinarian hold a PhD to undertake independent research. The extended period of time needed to become a biomedical investigator might significantly discourage students from pursuing this path. But despite the additional 4–5 years of effort, some DVMs pursue a PhD and postdoctoral training. Most typically, veterinarians entering the field of biomedical science and research do so through graduate degree(s) and postdoctoral training in a medical discipline. Some combine this with specialty training in clinical disciplines, such as laboratory animal medicine or veterinary pathology, leading to certification as a diplomate in the American College of Laboratory Animal Medicine (ACLAM) or the American College of Veterinary Pathologists (ACVP), respectively.

Data compiled by the NRC study *National Need and Priorities for Veterinarians in Biomedical Research* in 2004 indicate a strong but unfilled demand for veterinarians with proven research skills. The NRC report documented the rising number of position announcements for laboratory animal medicine veterinarians, which increased from less than 20 in 1995 to 50 in 2001. At the same time, animal use in the NIH grant portfolio is at an all-time high, a reflection of the continuing importance of animal based research (NRC, 2004b). Nearly 5,500 grants, or about 40 percent of all NIH competing grants, involved the use of live vertebrate animals (NRC, 2004b). However, NIH grants usually do not support animal disease research except as models for human disease. Veterinarians need to be trained in biomedical research to take active roles as principal investigators on NIH grants related to animal models for human disease and other grants for animal disease research, including investigations of the role of animals in zoonoses. The NRC report's review of Research Project (RO1) funded NIH grants in 2001 indicated that only 4.7 percent of NIH-funded competitive grants utilizing animals were awarded to veterinary principal investigators. The number of RO1s awarded to DVMs was small even during the period of doubling of the NIH budget (1997-2001): 76 RO1 awards to DVMs in 2001 (NRC, 2004b). The report concluded that the current number of veterinary investigators is not adequate to capitalize on the unique potential of comparative medicine to contribute to biomedical research.

The ACVP has provided further evidence of the future shortfall in biomedical scientists (ACVP, 2002). It studied the national needs for veterinary pathologists by surveying potential employers for the period 2002–2007 and compared this estimate to the expected output of trainees from existing training programs for the same period. It concluded there would be a shortfall of 336 pathologists or 50 percent of the predicted demand.

In summary, these facts point to a critical need for colleges of veterinary medicine to reexamine the nature of training provided to students relative to national needs. Although a more detailed examination of factors that impede veterinary students and veterinarians from pursuing research careers is beyond the scope of this report, these issues are the subject of a forthcoming NRC report entitled *Critical Needs for Research in Veterinary Science* (NRC, 2005).

Continuing Veterinary Medical Education (CVME)

In 2002, the AVMA Council on Education removed the Continuing Education Standard as essential for veterinary college accreditation. Hence this body no longer reviews college CVME programs. At present, no organization sets CVME national standards, as is the case for continuing medical education (Moore, 2003).

CVME is delivered by schools of veterinary medicine, various professional associations and societies, employers, and government agencies. Forty-one states, in one form or another, have mandated requirements for CVME to maintain licensure (Moore et al., 2003).

APHIS-VS administers the National Veterinary Accreditation Program (NVAP) (USDA-APHIS-VS, 2004) and plans to make regular CVME a mandatory requirement for the accreditation of private veterinary practitioners who wish to participate in federal and state regulatory programs (Torres and Bowman, 2002). It is anticipated that accreditation will be designated in two separate categories: one for companion animals and one for food-animals. Proposed rule changes are expected to be available for public comment in the winter of 2005 (Lawrence Miller, personal communication, June 2005). Currently, 80 percent of practicing veterinarians are accredited. The current accreditation program does not require veterinarians to maintain, through continuing education, their knowledge of foreign animal diseases. Under the proposed new program, foreign animal disease training will be available to complete CVME requirements to maintain accreditation status (Lawrence Miller, personal communication, June 2005).

Education and Training of Others on the Front Lines

Most animal handlers and others working and living with animals on a day-to-day basis are not health professionals and acquire their knowledge about animal disease through one or more means, such as from their veterinarian, employer, the Internet, industry magazines, commodity organizations, and extension programs offered by universities, government, or producer organizations. By definition, extension agencies are well positioned to take the initiative to provide appropriate training programs, but would probably require additional support to develop such instruction, given competing priorities and a challenging budgetary environment.

Wildlife agencies are expected to keep staff biologists and technicians adequately informed about disease issues. Hunters and naturalists can get information from societies dedicated to their interest or hobby through print, meetings, and the Internet.

AWARENESS OF THE ECONOMIC, SOCIAL, AND HUMAN HEALTH EFFECTS OF ANIMAL DISEASES

An outbreak of animal disease can have significant economic, social, and human health effects, although these effects vary considerably depending on the nature of the disease and the specific outbreak. Some animal diseases can have significant effects on markets. These include direct impacts on lost production and farm income, unintended costs to adjust from lost output, sector and community losses in welfare, and impacts on markets (prices) and trade. Consumers may lose confidence in the safety of meat and other food products, and this loss of confidence can contribute to a decrease in prices as well as lack of trust in public authorities. The potential for market and other impacts of an actual or threatened animal disease outbreak points to the importance of accurate and ongoing communication with consumers, producers, and the general public. Increasing dependence on trade can increase the volatility of prices. With the confirmed cases BSE in Canada in May 2003 and the Canadian-U.S. border closed to live cattle trade and only limited meat trade, U.S. beef prices rose by over 26 percent in 2003. After discovery of a BSE case in the United States in December 2003, U.S. beef prices fell by nearly 11 percent. The world beef trade declined by an estimated 2.5 percent in 2004 (Beghin et al., 2004). A recent review of studies of the economic impact of transboundary animal diseases indicates significant losses caused by the perceived threat of transboundary animal disease and control efforts. The studies include losses to Uruguay of added trade revenue estimated up to $90 million per year from the presence of FMD (1996) and losses in the

United Kingdom in 2000 related to BSE (lost trade, production, and other financial costs) of €5 billion (Otte et al., 2004). USDA estimates losses to the U.K. economy of $3.6–11.6 billion for FMD and $5.8 billion for BSE (USDA ERS, 2001). BSE is linked to variant Creutzfeldt-Jakob disease (vCJD) known to have caused 147 human deaths in the United Kingdom as of December 2004 (CJD Statistics, 2004).

In addition to known animal diseases from naturally occurring exposure is the added risk of disease that is spread with malicious intent (NRC, 2003a). Also, diseases associated with environmental disturbance or degradation are becoming more important. The effect of environmental contamination can affect domestic animal production, as well as the health of wildlife, and the value of hunting and fishing for recreation or livelihood.

3

Assessment of Current Framework: Case Studies

INTRODUCTION

In this chapter, a series of case studies are examined to assess the capabilities and limitations of the framework in preventing, detecting, and diagnosing animal diseases. The analysis of disease events that have occurred nationally and abroad provides useful information on the responsiveness of the framework as a whole and lays the groundwork for Chapter 4, which identifies gaps and opportunities to strengthen the framework. This chapter does not attempt to provide a comprehensive analysis of all animal diseases, but examines a "cafeteria-style" sample of diseases that could have potentially large economic, human, and/or animal health impact. Box 3-1 lists the animal diseases and disease categories selected for analysis. The list is not based on diseases that are the most problematic or prevalent in the United States. (For example, food-borne diseases caused by *Salmonella enteriditis* and *Escherichia coli* O157:H7 are not on the list; they pose greater concerns for the health of humans than for animals.) Instead, the animal diseases or disease scenarios described here, from acute to chronic, endemic to exotic, naturally occurring to intentionally introduced, were selected to consider the breadth of issues that must be addressed by an inclusive infrastructure capable of detecting, diagnosing, and preventing a wide variety of events affecting animal and human health. The diseases selected involve each of the major animal types, namely food-animals, wildlife, and companion animals.

FOREIGN ANIMAL DISEASES: EXOTIC NEWCASTLE DISEASE AND FOOT-AND-MOUTH DISEASE

Exotic Newcastle Disease

Exotic Newcastle disease (END) is a contagious and fatal disease affecting all species of birds. Previously known as velogenic viscerotropic Newcastle disease (VVND), END is one of the most infectious diseases of poultry worldwide. A death rate of nearly 100 percent can occur in unvaccinated poultry flocks. The virus is so virulent that many birds die prior to showing clinical signs, and END infection can have high mortality even in vaccinated birds (University of Georgia, 2003).

END is classified as a foreign animal disease in the United States, historically causing severe economic losses when commercial poultry industries become infected, as occurred in a major outbreak of END in southern California in 1971. The disease threatened not only California poultry production, but it also had a significant economic impact on the entire U.S. poultry and egg industry. In all, 1,341 infected flocks were identified and 11.9 million birds were destroyed over a multiyear disease control effort. Disease eradication cost taxpayers $56 million (over $250 million in 2003 dollars), severely disrupted the operations of many producers, and increased the price of poultry and poultry products to U.S. consumers (Utterback, 1973; Davidson-York et al., 1998). It took 3 years to fully eradicate the disease, and nearly two decades before another outbreak of END occurred in U.S. commercial poultry. In the early 1990s, over 26,000 commercial turkeys were destroyed in North Dakota following detection of END. The virus is believed to have been transmitted to the turkeys from cormorants or other free-ranging birds. Hundreds of cormorants had previously died at a lake not far from the turkeys, in an outbreak that is believed to be the first documented Newcastle-related die-off of wild birds in the United States (Meteyer et al., 1997). Though END virus has not been detected in commercial birds in the United States since then, it is now known to exist in free-ranging wild birds, as well as in psitticine species. A variety of psitticine species enter the United States through the pet bird trade, generally traveling through USDA quarantine stations; however, illegal movements across U.S. borders also occur. END is detected nearly every year in California, primarily in psitticine and free-flying wild-bird species; however, in 1998, END was detected in urban gaming chickens in the state (Crespo et al., 1999). Subsequent to the 1971 outbreak, the presence of END has been detected numerous times through case submissions to the state's diagnostic laboratory (passive surveillance), confirmed as END by the federal laboratory system, and rapidly

BOX 3-1
Animal Diseases Addressed in This Chapter

Foreign Animal Diseases. Important transmissible livestock or poultry diseases that are largely absent from the United States and its territories and that have the potential to cause significant health or economic impact should the causative agent be introduced. Foreign animal diseases discussed in this chapter include:

- **Exotic Newcastle disease (END)**
- **Foot-and-mouth disease (FMD)**

Recently Emergent Diseases. Infectious diseases for which the risk in animals has increased in the past two decades or threatens to increase in the near future. These diseases include:

A. New infections resulting from changes or evolution of existing organisms or newly infectious particles (such as prions)
B. Known infections spreading to new geographic areas or populations
C. Previously unrecognized infections emerging in new geographic areas and human populations due to changing technologies and behaviors
D. Old infections reemerging as a result of antimicrobial resistance in known agents or breakdowns in animal disease control measures.

The recently emergent diseases addressed in this chapter are:

- **Monkeypox**
- **Bovine spongiform encephalopathy (BSE)**

Previously Unknown Agents. Pathogens previously unrecognized that have recently (within the past decade) been transmitted from animals to humans. Included for discussion in this chapter:

- **Severe acute respiratory syndrome (SARS) coronavirus**

eliminated by state regulatory authorities prior to spread of the disease (Molenda, 2003).

Detection and Diagnostic Methods

Despite the recognized and significant economic impacts of END introduction into the U.S. commercial poultry industry and the repeatedly observed risk of reintroduction in California, surveillance, detection, and diagnostic approaches were little changed in 2002 from those used in 1971. The accepted diagnostic standard was virus isolation in embryonated

Endemic Diseases. Animal-borne diseases that are native to or commonly found in the United States. Examples addressed here include:

- **Avian influenza**
- **Chronic wasting disease**
- **West Nile virus**

While the committee recognizes that at one time these agents may have been considered as newly emergent, each of them has now become firmly established in North America and is considered endemic for the purposes of this report.

Novel Naturally Occurring Pathogens. Organisms previously unreported or infrequently associated with being a primary pathogen in a given host species. Novel naturally occurring pathogens may contain new genomic elements acquired through natural processes and not as the result of in vitro insertion.

Bioengineered Animal Pathogens. Organisms containing genomic elements that were acquired in vitro.

Diseases of Toxicological Origin. Diseases caused by exposure to toxic substance(s), including drug residues, in a concentration that alone or in combination meets either of the following criteria: (1) the animal(s) affected is/are a potential source of toxicological contamination to humans or other animals and/or (2) the source of the toxicological agent or exposure is potentially hazardous to humans or other animals.

eggs, a process requiring 2 to 7 days, followed by pathogenicity testing of the isolated virus by inoculation into chickens or direct nucleic acid sequence analysis of the virus' pathogenicity marker. Isolation and characterization of the virus requires several days to several weeks, depending on the availability of eggs and experimental birds, access to containment and/or sequencing facilities, and technical resources at the federal laboratory. Though state and university veterinary diagnostic laboratories typically have virus isolation facilities with trained technical staff, consideration had not been given to using these resources; instead, the existing paradigm was for the federal laboratory to perform foreign animal disease testing. National technical training and proficiency evaluation for the isolation and characterization of foreign animal diseases, including END virus, did not exist, which limited possibilities for providing surge capacity needed in the face of an END outbreak. Proven technology that would allow surge capacity in the form of rapid and sensitive diagnostic assays had not yet been directed toward END detection, primarily because the U.S. Department of Agriculture (USDA) Cooperative State Research, Education, and Extension Service (CSREES), the major funding agency for animal health, discouraged allocation of competitive research dollars for projects solely targeting the development and validation of veterinary diagnostic assays.

However, in response to heightened biothreat awareness in early 2002, the USDA, in cooperation with the Department of Homeland Security, created a list of eight high-risk agriculture pathogens (USAHA, 2003a). Included in the $14 million funding allocated to the USDA Agricultural Research Service (ARS) for developing rapid diagnostics for the high-risk agricultural pathogens was $2.8 million for two poultry pathogens included in the list: highly pathogenic avian influenza (AI) virus and END virus (USDA, 2002f; USDA ARS, 2002). Months later, prior to the development or availability of rapid detection assays in the United States (though rapid END diagnostic approaches were documented in the international literature), END was again found in game fowls in southern California (Nolen, 2002). The END outbreak illustrates the following findings:

- The animal health infrastructure lacked an analysis system for anticipating challenges to animal agriculture and a system for providing appropriate intervention or rapid detection strategies despite acknowledged risks of introduction of a high consequence pathogen into the United States.
- The existing infrastructure did not support timely development, validation, and implementation of state-of-the-art technologies for prevention, detection, and diagnosis of recognized and economically threatening pathogens.

The 2002 END Outbreak

The timing and movement of the END outbreak in 2002–2003 followed a pattern eerily similar to the 1971 outbreak (see Table 3-1). In late September 2002, a game chicken was presented to the state's animal health laboratory by a private veterinary practitioner on behalf of a southern California game fowl owner that had lost 200 birds (90 percent mortality) over a 5-day period. Two days later a veterinarian in a neighboring county contacted the state laboratory to report high mortality in a small backyard flock of laying hens. Within the previous 6 months, two unrelated companion animal bird submissions with confirmed END infections had been traced to origins in southern California. In all cases the laboratory suspected END virus, and samples were transported to the federal laboratory in Ames, Iowa, as required for confirmation of END (Humanitarian Resource Institute, 2004). The END viruses isolated in the spring and fall of 2002 had identical genomic sequences, suggesting the virus may have entered bird populations in southern California at least 6 months before the declared outbreak. By the time the initial game fowl cases were detected by passive surveillance in late September, the disease had spread throughout the urban population of game and noncommercial poultry in southern California. The size and significance of the urban poultry population had clearly gone unrecognized, and the social and cultural barriers to effective surveillance within that population had not been addressed. Within the first week of the outbreak response alone, more than 5,000 noncommercial birds were depopulated and 30 backyard flocks placed under quarantine in a three-county area. Ultimately, nearly 300,000 premises were visited during the outbreak, and 90,000 of them had avian species, primarily poultry. Though trade partners had been notified by USDA of the END detected in game chickens on October 1, 2002, a federal emergency was not declared until January 6, 2003, by which time virus had been detected in 5 of the ultimate 22 infected commercial poultry flocks (USDA APHIS, 2003b).

Requests to the European Union for regionalization to protect U.S. trade and questions of federal and state authorities had, however, been initiated by USDA in late October and November (Rob Werge, personal communication, 2004). Heightened awareness of the disease resulted in the detection of END in neighboring Nevada game chickens in mid-January 2003, Arizona game chickens in early February, and an isolated incident of an unrelated END virus in Texas game chickens in early April. By the time the final END positive bird was detected 9 months later, 22 commercial premises and a total of 3.21 million birds had been depopulated at a cost of more than $160 million in federal control efforts. Up to 71 percent of USDA veterinary services staff—and a total of 7,690 state and federal employees—were recruited into the eradication effort (USDA APHIS, 2004a). Based on this review, the committee found:

- The existing animal health infrastructure was designed to detect and respond to disease in commercial agriculture production systems, and was not appropriate for nontraditional species, management, or environments, effectively delaying both detection and response activities.
- The lack of adequate surveillance for END, a foreign animal disease already known to enter the country periodically, allowed the virus to

TABLE 3-1 Timeline of 2002-2003 Exotic Newcastle Disease (END) Outbreak in Southern California

2002	March	• Two unrelated companion animal birds diagnosed with END. Federal efforts trace origin of birds to Southern California.
	September 25	• Index case END outbreak game fowl submitted to state diagnostic laboratory by private practitioner.
	September 27	• Second case in backyard chickens submitted to state laboratory. • Task force formed. Urban door-to-door disease eradication efforts begin.
	October 1	• National Veterinary Services Laboratories (NVSL) confirms END virus. U.S. Department of Agriculture (USDA) notifies state veterinarians and trade partners. EU bans import of U.S. poultry. • USDA transfers $7.4 million in funds from Virginia LPAI to END ($5 million for state, $2.4 million for USDA VS). • Initial development of rapid diagnostic assay begins.
	November	• Prototype Real-time Reverse Transcriptase (RRT) Polymerase Chain Reaction (PCR) used in conjunction with virus isolation by state laboratory to detect END.
	December 23	• First detection of END in a commercial flock, presumptive diagnosis by state laboratory using virus isolation and RRT PCR, officially confirmed December 21 by USDA testing. • Secretary of Agriculture approves $121.8M fund request for END control.
2003	January	• USDA declares extraordinary emergency. • Prototype RRT PCR used in federal laboratory. • END detected in Nevada game fowls (January 16, 2003). Nevada Task Force established. USDA declares extraordinary emergency for Nevada (January 17, 2003). • E.U. and Mexico agree to regionalize United States, restricting trade only with California, Nevada, and Arizona. • Arizona Task Force established due to proximity to California quarantine zones.
	February 4	• Arizona game fowl confirmed positive for END.
	February 7	• USDA declares extraordinary emergency for Arizona. • Virus isolation (egg inoculation) reaches peak laboratory capacity at ~4,000 samples per month.
	March	• Last detection of END in a commercial poultry flock.

April 9	• END detected in Texan game fowl.
April 10	• USDA declares extraordinary emergency for Texas and border state New Mexico. • USDA reports genomic sequence of Texas isolate differs from outbreak virus indicating introduction not due to spread from California, Nevada, and Arizona. • USDA officially validates USDA single-tube END RRT PCR. Test turn-around is 4–24 hours, laboratory capacity ~184 tests per day based on 3 cyclers and 3 technicians.
May	• Final positive noncommercial bird detected. • State laboratory initiates use of modified high-throughput RRT PCR. Test turn-around is 4–24 hours, laboratory capacity ~1900 samples per day. • USDA lifts quarantines from all but original infection sites in Nevada and Arizona.
June	• USDA lifts quarantine in Texas and New Mexico except for area around original positive premise.
July	• Quarantines lifted from Arizona, Nevada, Texas, reduced areas in California.
August	• E.U. lifts trade restrictions except for Southern Califonia, and original infection sites in Arizona, Nevada, and Texas. • California and Mexico sign agreement for regional plan for prevention and mitigation of future END outbreaks.
September	• USDA approves $9.476 million for END surveillance. • USDA lifts California quarantine. Surveillance efforts directed toward avian health and mitigation continue. • Mexico and Canada lift END-related trade restrictions.

Final totals

- 19,146 premises quarantined
- 932 confirmed infected premises identified
- 3.21 million birds depopulated in four states
- $160 million in control costs
- 7670 state and federal employees on the END task force

---

spread and become established in a relatively large animal population before detection.

In the absence of available rapid detection and diagnostic assays for END, the USDA and the state diagnostic laboratory, in a largely uncoordinated effort, initiated development of a molecular-based diagnostic approach to reduce from days to hours the time needed to obtain an END diagnosis. The federal approach was based on past experiences and successes with flock-based detection, while the state responded to the character of the current outbreak by focusing on high-throughput capacity and reliable detection in individual birds. Both groups used the established technology of real-time polymerase chain reaction (PCR). The state laboratory lacked significant fiscal resources specifically allocated for assay development and so relied on partnerships and collaborations with other federal agencies (such as the Department of Energy) and with commercial biotechnology equipment and reagent suppliers. The use of the extensive network of expertise located outside of the federal (and state) system allowed for timely development of a more cost-effective and rapid approach to detection and diagnosis of END, which was ultimately used on more than 81,000 samples during detection and control efforts by the state laboratory. The combination of real-time PCR with the high-throughput approach allowed a 10-fold increase in workload to more than 1,500 samples tested daily with results available within 4 to 24 hours, and is credited with supporting rapid and effective testing for disease eradication. The final END quarantines were lifted within 11 months of initial END detection in game fowls, despite earlier USDA Animal and Plant Health Inspection Service (APHIS) projections of a 3-year disease control effort (USDA APHIS, 2003c). The committee found the following:

- Private industry, local and regional resources, and the willingness to capitalize on expertise located outside the centralized federal animal health system allowed a more timely, cost-effective, and reliable assay to be developed, validated, and implemented for disease detection and control.

Foot-and-Mouth Disease

Foot-and-mouth disease (FMD) is a highly contagious viral disease of cattle, swine, and other cloven-hoofed species including sheep, goats, and deer. The disease is characterized by fever and blister-like lesions followed by erosions on the tongue and the lips, in the mouth, on the teats, and between the hooves. For some strains of the virus and host species, clinical signs of infection can be minimal or go clinically unrecognized. Most affected animals recover, but the disease can leave them debilitated and livestock herds can experience severe losses in production of meat and milk, providing the economic justification for including FMD virus among the OIE List A diseases (OIE, 2003). Pigs amplify most strains of FMD virus to high concentrations, so they transmit the disease readily, while cattle are generally considered the species most susceptible to infection. The virus can be transmitted readily to susceptible animals either by ingestion or inhalation of the virus from contagious animals or innate objects, such as contaminated vehicles, clothing, or feed or water. The virus is well known for its potential to spread widely and rapidly in the absence of obvious clinical signs that would trigger early detection and appropriate control measures.

FMDV in the United Kingdom

The United States has not had an incursion of FMD virus (FMDV) since 1929, but the devastating outbreak of FMD in the United Kingdom in 2001 (Box 3-2) has provided lessons about prevention, detection, and control of the disease in the United States. As in the United Kingdom, the United States does not permit the use of FMDV vaccine, creating a national population of FMDV-susceptible animals. Moreover, the United States has a large wildlife population—including feral swine, deer, and other susceptible cloven-hoofed animals—for which timely detection and prevention would be difficult, if not impossible. Establishment of infection in susceptible wildlife, such as feral swine, could result in widespread dissemination of the disease throughout the country. Prevention in both countries was and continues to be heavily reliant on federal policies restricting trade in animals and animal products from FMDV-endemic countries. Despite such policies, in early 2001 the FMDV entered the United Kingdom, most probably through an illegally imported meat product. By the time the disease was detected several weeks later, it had spread throughout the country and to as many as 79 premises primarily through animal movement (Mansley et al., 2003). Disease entry through import, either intentional or unintentional, is a similar risk for the United States, where a very small percentage of cargo and baggage is inspected. The USDA Safeguarding Review cites that 489 million passengers and pedestrians and 140 million conveyances crossed U.S. borders in 2000, and the review predicted this number to double in 2009 (NASDARF, 2001). In addition, approximately 38,000 animals were imported daily into the United States in 2000. The committee found that:

- FMD prevention, and disease prevention in general, through exclusion of infected animals and animal products cannot be relied on as infallible and would require a significantly more effective infrastructure than currently exists at U.S. borders and ports of entry.

The lack of early detection following FMD virus introduction in the United Kingdom was responsible for the widespread dissemination of disease throughout the country and into neighboring countries (Haydon et al., 2004). Standard methods for testing clinical material (lesion swabs,

BOX 3-2
Foot-and-Mouth Disease Epidemic in Great Britain in 2001

On February 19, 2001, a routine inspection at an abattoir near London revealed "highly suspicious" signs of foot-and-mouth disease in 27 pigs. The Ministry of Agriculture confirmed the outbreak and the next day set up a 5-mile exclusion zone around the abattoir. With increasing numbers of FMD cases reported on cattle and sheep farms 5 days after the initial case, the government announced plans to slaughter pigs, sheep, and cattle in affected areas in an attempt to eliminate the disease. As the outbreak continued through the end of the month, the ban on movement of livestock was extended. By early March, neighboring countries had begun investigating their own suspected cases of FMD and enhanced precautionary measures were initiated to prevent FMD from entering their countries. The epidemic, however, extended beyond England to other European countries, with Scotland, Northern Ireland, France, Belgium, Denmark, Germany, and the Netherlands responding with programs to destroy animals in affected areas.

At a meeting of European ministers on March 6, a proposal was made to extend the ban on British livestock exports until March 27. Veterinary experts recommended against mass vaccination, and the E.U. agriculture ministers concurred with their advice. Despite extensive efforts, the number of new unconfirmed cases reached 1,000 by the beginning of April. On April 26, the government announced a change in policy, ending the practice of slaughtering healthy, unaffected livestock on farms neighboring farms with animals showing suspicious signs. By May 8, restrictions on livestock movement were eased across the European Union.

The British government killed 6.5 million animals during the epidemic: about 4 million for disease control and an additional 2.5 million for reasons of animal welfare. The epidemic lasted 214 days and involved over 10,000 herds and flocks. Annual festivals and international sporting events were cancelled due to the epidemic and tourism declined substantially. The epidemic incurred losses to agriculture and tourism estimated to be at least £6.3 billion.

SOURCES: Thompson et al., 2002; Haydon et al., 2004.

fluids, blood) from suspect animals, including virus isolation in cell culture and antibody detection by serum testing, were available in the United Kingdom and used with reportedly high accuracy. Had the government invested years earlier in the development of accurate and rapid real-time virus detection assays, it would have been very difficult for the government then to ignore the need for diagnostic testing before animals were destroyed. Traditional FMDV control methods, targeting the killing of animals from infected premises and epidemiologically determined dangerous contact premises, were used through late March, after which time additional control measures were introduced. These measures included depopulation based only on suspicion of infection; destroying sheep, goats, and pigs within 3 kilometers of infected premises in some counties; and destroying animals on all premises contiguous to an infected premise within 48 hours regardless of health status of the animals. The 48-hour depopulation policy was and remains a controversial component of the U.K. control effort, and therefore was not consistently accepted. The policy, developed in large part based on computer model simulations of hypothetical disease transmission, has in retrospect been credited with the large number of noninfected animals destroyed during the 2001 epidemic and with the public's negative response to the highly visible control efforts (Haydon et al., 2004). From the published lessons identified and formal recommendations in commissioned reports (National Audit Office, 2001; Royal Society, 2002), it can be concluded that:

- The lack of early detection allowed FMD to become widespread in the United Kingdom.
- Outbreak planning with established, scientifically consistent policies and protocols defined prior to the outbreak or disease event are necessary for effective prevention, diagnosis, and response.

FMDV in Other Countries

Unlike the United Kingdom, the Netherlands chose to respond to the related 2001 incursion of FMDV into its herds with an emergency vaccination program (Tomasson et al., 2002; Bouma et al., 2003). The program was successful in the Netherlands and is cited as justification for the emergency use of FMDV vaccination during an outbreak, despite the after-affects of restricted trade when all vaccinated animals are not subsequently destroyed (Haydon et al., 2004). The criticisms of vaccinating in the face of an outbreak include the current lack of a validated assay or technology that would allow for the differentiation of animals exposed to FMD vaccine from those animals exposed to the live virus. The potential for an exposed and vaccinated animal to become a subclinical FMD virus carrier, capable of disease spread, is a significant concern for trade partners in FMD-free countries following an FMD outbreak in vaccinating countries. Technologies utilizing animal serum to test for their response to portions of the replicating FMD virus, termed nonstructural protein assays, have been developed in recent years but have not yet been evaluated or approved for use in the United States. Likewise, technologies to produce effective vaccines that can rapidly and effectively protect an animal and also allow paired diagnostic tests to distinguish vaccinated from exposed animals (marker vaccines and diagnostics) exist for other animal diseases but are not yet developed or readily available for FMD control. The lessons reported from the 2001 U.K. FMD outbreak indicate that there is an immediate and ongoing need to provide for the development and critical evaluation of advancing technologies for vaccines, as well as detection and diagnostic assays for disease prevention, detection, and control.

RECENTLY EMERGENT DISEASES IN NORTH AMERICA: MONKEYPOX AND BOVINE SPONGIFORM ENCEPHALOPATHY

Monkeypox

Monkeypox is a rare viral disease that is found mostly in the rain forest countries of Central and West Africa. The disease is called "monkeypox" because it was first discovered in laboratory monkeys in 1958. Blood tests of animals in Africa later found evidence that monkeypox is primarily an infection of rodent species. The virus that causes monkeypox was recovered from an African squirrel, which may be the natural host. Laboratory studies showed that the virus can also infect rats, mice, and rabbits (Khodakevich et al., 1986; Hutin et al., 2001).

In 1970, monkeypox was identified as the cause of a rash illness in humans in remote African locations (Landyl et al., 1972; CDC, 2003b). Interestingly, in retrospect some monkeypox may have been misdiagnosed in humans prior to this time as mild smallpox but was easily identified as a separate disease after smallpox was eradicated (Ogden, 1987). In early June 2003, monkeypox was reported among several residents in the United States who became ill after having contact with sick companion animal prairie dogs. (See Box 3-3 for a description of the case.) This is the first evidence of monkeypox in the United States.

Prevention

There was no formal provision for monitoring monkeypox in these animals by an appropriately trained health professional at the point of origin in Ghana, at the importer, or from the importer on into the United States. Because of the lack of records, 178 (23 percent) of the original 762 African rodents could not be traced beyond the Texas importer (CDC, 2003e). Furthermore, there were no health examinations, certificates, or individual animal identification required for the prairie dogs exposed to

BOX 3-3
Recent Emergence of Monkeypox in the United States

On April 9, 2003, a shipment of 762 exotic rodents originating in Accra, Ghana, reached the United States. That shipment contained giant Gambian pouched rats (50 animals), rope squirrels (53), brushtail porcupines (2), tree squirrels (47), striped mice (100), and dormice (510). Accompanying these animals to Texas was an unexpected virus that eventually found its way into at least two other animal species in the United States (prairie dogs and humans) and spread to at least six other states. That unexpected agent, previously unseen in the United States, was a member of the orthopoxvirus group known as monkeypox (CDC, 2003b). It brought a scare to a public health and homeland security infrastructure, already in a state of heightened awareness for smallpox, and challenged the ability to address an emergent health threat in the United States that did not conveniently fall under the domain of any single federal agency.

In mid-May 2003, the first human cases of a febrile vesicular rash in the United States were examined by physicians in Illinois and Wisconsin. By June 10, a total of 53 cases were being investigated, 51 of which reported contact with a companion animal prairie dog. The Marshfield Clinic in Marshfield, Wisconsin, isolated and identified a virus from vesicular lesions of a human patient and from lymph nodes of the patient's companion animal prairie dog. That virus, when examined by electron microscopy, resembled a poxvirus. The U.S. Centers for Disease Control and Prevention (CDC) subsequently identified it as monkeypox (CDC, 2003b). Spread of this poxvirus had peaked by early June, but in total over 70 cases from six states—Illinois, Wisconsin, Indiana, Kansas, Missouri, and Ohio—were reported over an approximately 3-month period. In Indiana, 28 children were exposed to a companion animal prairie dog, and seven became ill following this exposure (Langkop et al., 2003).

How did this virus make its way from exotic rodents in Ghana to a classroom in the heartland of the United States? Did it cause clinical signs in animals? Knowing that rodents in Africa carry monkeypox, why were these animals allowed into the United States, or at least not tested for the virus before entry? Who had responsibility for surveillance, identification, and response to this foreign zoonotic agent in exotic companion animals? These were some questions the committee asked while studying the monkeypox outbreak.

the monkeypox, which were distributed to eight states through "swap meets" where exotic animal aficionados gather to trade specimens.

Poor or no sales records are kept at swap meets, all of which complicated efforts to trace back and trace forward animals from the original shipment and exposure contacts. At the time that the first human cases were being examined, the extent of the problem in animals was not understood, nor was it known that exposure to prairie dogs (and only prairie dogs) would turn out to be central to human cases. Also not known, at that time or at present, was the susceptibility of various animal species to monkeypox infection. Vendors of exotic companion animals often keep an impressive variety of species, many of which could have been susceptible. Once a definitive diagnosis was made, the Centers for Disease Control and Prevention (CDC) took a lead role in notifying regulatory officials of the outbreak.

On June 11, 2003, CDC and the Food and Drug Administration (FDA) issued a joint order that announced an immediate embargo on the importation of rodents from Africa and banned any sale, offering for distribution, transport, or release into the environment of prairie dogs and six genera of African rodents potentially involved in the spread of monkeypox in the United States (CDC, 2003b). CDC has jurisdiction over the importation section of the rule, while FDA has jurisdiction over movement of animals between and within states. On November 4, 2003, the joint order was replaced by an interim final rule that maintains the bans on importation of these rodents and their sale or distribution. These actions were taken by the Department of Health and Human Services under the authority granted in Section 361 of the Public Health Service Act (42 U.S.C. 264). Section 361 grants the Secretary of Health and Human Services the authority to make and enforce regulations to prevent the introduction, transmission, or spread of communicable disease from foreign countries into the United States or from one state to another. States are free, within their legal authority, to enact other regulations as long as those regulations do not conflict with the interim final rule. Enforcement of this rule relies on the CDC and FDA working collaboratively with other federal and state agencies. Many federal, state, and local agencies have authorities related to the animals involved, including the USDA and state departments of agriculture, which oversee the trade in these animals within the United States; and the Bureau of Customs and Border Protection of the Department of Homeland Security and the Fish and Wildlife Service of the Department of the Interior, which have statutory authority for enforcing importation embargos. The interim final rule addresses many of the issues surrounding importation and movement of exotic rodents into and within the United States. However, it does not ban the importation of all exotic animals.

The monkeypox outbreak revealed that:

- The infrastructure that exists for preventing animal disease outbreaks is focused primarily on livestock, including poultry and farmed aquatic animals. There is no equivalent federal responsibility and only an informal federal animal health infrastructure for addressing a zoonotic disease outbreak transmitted by a nonlivestock species.

Diagnostic Laboratory Capacity

State and academic veterinary diagnostic laboratories play a central role in diagnosing diseases such as monkeypox. As the monkeypox outbreak illustrates, the broad capabilities that exist in state and academic veterinary diagnostic laboratories and other local animal health infrastructure are currently underutilized and underconsulted by federal agencies and national organizations and by the public health community.

The monkeypox outbreak also points to weaknesses in the veterinary laboratory infrastructure in the United States. Far too few biosafety level 3 (BSL-3) laboratories and ABSL-3 animal facilities exist in the state laboratory system, universities, and industries (AAVMC, personal communication and informal survey, 2004). Rapid assays for zoonotic agents, whether endemic or exotic, have not been validated in animals. Assays validated for identification of exotic or bioterrorist agents in human samples (such as the non-variola rapid real-time polymerase chain reaction (PCR) assay used for diagnosis of monkeypox) do not have animal species or matrix controls. Further complicating this issue is that no federal agency has a mandate to develop and validate these assays. USDA is fully committed with livestock disease assay development and validation, and CDC is focused on development of additional assays for the broad array of high consequence pathogens affecting humans. The diagnosis of overlap agents in animals has fallen through the cracks. Good laboratory practice includes training and proficiency testing of laboratory staff in use of equipment and specific protocols, but neither the Laboratory Response Network for bioterrorism nor the National Animal Health Laboratory Network currently has responsibility for ensuring that these tenets of a quality system are in place in veterinary diagnostic laboratories. In summary, the monkeypox outbreak revealed significant gaps in prevention, problems with responsibilities, and coordination of response and in laboratory capacity, especially concerning delays in the development and validation of diagnostic assays.

Bovine Spongiform Encephalopathy

The diagnosis of bovine spongiform encephalopathy (BSE) in Canada and the United States in 2003 (see Box 3-4) and in 2005 carried with it a message that North America was not immune to the socioeconomic effects of what is commonly known as mad cow disease. Disruptions in the

BOX 3-4
Single Case of Bovine Spongiform Encephalopathy (BSE) in Washington State: An Unexpected Opportunity for Insight into Our Framework for Preventing, Detecting, and Diagnosing Animal Diseases

On the morning of December 25, 2003, the BSE World Reference Laboratory in Weybridge, England, confirmed USDA's December 23 preliminary diagnosis of BSE in a single nonambulatory dairy cow that had been slaughtered on December 9 at Vern's Moses Lake Meats in Washington State. USDA and Canadian officials worked together to confirm the identification of this cow through DNA testing and to establish that the animal was imported from Canada.

BSE (or "mad cow" disease) is a neurodegenerative disease transmitted to cattle through contaminated feed. It has an incubation period of 4–6 years. It is caused by an aberrant form of a protein called a prion and is in the family of diseases—all caused by prions—referred to as transmissible spongiform encephalopathies, or TSEs. The prion is an abnormally folded version of a normal cellular protein. The abnormal conformation results in a phenotype that is highly resistant to degradation and can persist in an infectious form during the rendering of contaminated bovine by-products for animal feeds, and the preparation of other products such as cosmetics and drugs using ingredients derived from cattle.

Unlike other agents (such as FMD virus), the prion is not contagious. Although it is considered "infectious," it is not spread directly from one animal to another. It carries a low risk of spreading to animals in the United States (Harvard risk assessment). The overall public health risk of develop-

supply of meat can shake consumer confidence, resulting in reduced demand, and can significantly disrupt trade of meat and meat products for a prolonged period. Establishing countrywide disease-free status once a case is diagnosed can be extremely difficult. According to a panel of experts from the European Association for Animal Production, the estimated total cost of BSE in Europe is €92 billion, nearly $115 billion dollars (EAAP, 2003). It had been estimated that a single case of BSE in either Canada or the United States would cost their respective beef industries $3.3 billion CAD and $6 billion USD, respectively (CBC News, 2003; Presley, 2004).

The onset of BSE in Great Britain led the United States to carry out an extensive analysis and forge policies based on risk factors associated with the disease, even though the disease was not present; this marked a significant departure from the past (USDA APHIS-VS, 1991). Trade in animal feed has been extensive in North America. Rendered by-products ing variant Creutzfeldt-Jakob disease (vCJD, the human variant of the disease, which is acquired through consumption of prion-contaminated meats) from a few cases in the bovine population or through blood transfusion is extremely, almost infinitesimally small.

At present, immunohistochemistry and immunoblot are widely considered in the international community the two gold standards to test for BSE.

The key to prevention is to ensure that high-risk materials from cattle are not incorporated into the feed supply. Enforcement is critical to the success of this approach. When enforcement cannot be guaranteed, a complete ban on feeding of ruminant by-products may be necessary.

In this case, the USDA Food Safety and Inspection Service recalled over 10,000 pounds of meat to prevent human food contamination. The recall involved by-products from 20 BSE-infected cattle, including 2,000 tons of potentially infectious feed, which had already been processed and exported to foreign ports. Over 700 animals were slaughtered during the traceback and traceforward investigation, while the U.S. beef industry continued to see a loss of export markets.

Lessons identified from the BSE experience included the following:

- The U.S. animal health community realized that BSE can no longer be considered a problem only for other nations.
- In contrast to traceback required for a highly contagious diseases like FMD and classical swine fever, comprehensive tracing required for a disease like BSE, with such a prolonged incubation period and likely exposure as a young calf, was nearly impossible within the current U.S. system of animal tracking and identification.

from the United Kingdom were freely imported into North America prior to an understanding of the potential of these products to transmit BSE. In addition, U.S. and Canadian restrictions that ban feeding of ruminant by-products to other ruminants were not implemented until 1997, and even then compliance, at least in the United States, may not have been optimal. Thus, the advanced age (6½ years) of the BSE-infected animal in the United States placed her, and her birth cohorts, at risk of exposure to BSE-contaminated ruminant by-products as a calf.

The BSE prion (PrPres) concentrates almost exclusively in nervous tissue in cattle and is found in highest concentrations in the brain, eyeballs, spinal cord, and dorsal root ganglia. In younger animals, the distal ileum, or last segment of the small intestine, can also harbor PrPres. This is the basis for concentrating on control of these so-called specified risk materials (SRMs). Appropriate quality control can ensure that these materials are not used for ruminant feed or human food. An added level of assurance can be provided by a complete ban on feeding of ruminant products back to ruminants, while complete assurance can only be provided by a complete ban on the use of ruminant by-products. In examining the 2003 BSE case, the committee concluded the following:

- The key to preventing an accidental introduction of BSE into the United States, as well as preventing subsequent transmission to other animals and humans, is recognition of the sources of infection and means of spread. Control of import, production, and distribution of ruminant by-products for feed, food, drugs, and cosmetics is essential.
- A risk-based approach is best used to determine what level of control should be implemented. The World Animal Health Organization provides a model for risk analysis.

In the United States, several additional control steps were taken following the discovery of the BSE-infected animal in December 2003. They included prohibition of all nonambulatory, or "downer," cattle and SRMs (skull, brain, trigeminal ganglia, eyes, vertebral column, spinal cord, and dorsal root ganglia of cattle over 30 months of age and a portion of the small intestine of cattle of all ages) from the human food chain; prohibition of a meat label for dorsal root ganglia that might be present in products obtained through advanced meat recovery processes; prohibition of air injection stunning of cattle at slaughter; prohibition of mechanically separated meat in human food; holding product from BSE-tested animals until a final diagnosis has been made; and immediate implementation of a national animal identification system. These new rules were published in the Federal Register on January 12, 2004.

BSE does not generate a typical host immune response. To date there has been no demonstration of antibodies generated to the abnormal prion variant in any affected species. Thus, traditional methods of infectious disease control by vaccination hold little if any promise for BSE. Perhaps it is possible to stop infection of cattle by means other than preventing exposure. To address this possibility, further research on the process of uptake and dissemination of the abnormal prion and conversion of normal prion to the abnormal variant is necessary. With a better understanding of the pathogenesis of these unique agents may come novel methods of prevention through blocking of transmission or disease progression, which leads to the following conclusion:

- Early detection of BSE relies on recognition of clinical signs and testing of the appropriate, high-risk population of animals.

While a thorough discussion of surveillance is not within the scope of this report, the committee recognizes that surveillance of high-risk animals (nonambulatory, with or without central nervous system signs) provides the highest sensitivity of early detection of BSE. The surveillance program implemented by the United States in June 2004 is designed to detect a level of BSE as low as five cases in the U.S. high-risk group.

At the time of publication of this report, a second BSE case had just been confirmed in the United States. A downer cow initially tested "inconclusive" for BSE via enzyme-linked immunosorbent assay (ELISA) rapid screening and immunohistochemical (IHC) tests in November 2004, but a later immunoblot (Western blot) test resulted in a "weak positive." The sample was retested in June 2005 at the world reference laboratory for BSE in Weybridge, England, and was confirmed positive using a combination of rapid, Western blot, and IHC tests (USDA, 2005). This case raises questions about the type and accuracy of diagnostic tests used by USDA to confirm an initial reactor in the ELISA assay.

Diagnosis of BSE relies on the use of ELISA assays, Western blots, or IHC tests. Each of these assays requires sampling of brain from dead animals and demonstrates through antibody binding the presence of the abnormal prion, PrPres. All of the currently approved assays have excellent sensitivity in older animals (30 months and older). However, in younger animals, the prion is either not present or is present at such low levels in the target tissues that it cannot be detected. The decision on which assay is used depends upon test purpose and the fitness of each assay for that purpose. High throughput formats are most commonly used for surveillance purposes to facilitate testing of large numbers of animals. Immunohistochemistry, in which the presence of the abnormal prion is visualized under the microscope in a section of brain by a trained pathologist, and Western blots, in which the abnormal prion in the brain can be visualized and its approximate molecular weight determined after separation from other proteins in the brain based on its size and resistance to enzymes, are widely considered gold standard tests. Immunohistochemistry, however, cannot be used for high throughput testing, in which rapid turnaround is required, as results typically are not available for 2–4 days, leading to this lesson learned:

- There is no available method for diagnosing BSE in young calves, when infection first occurs, or in live animals.

At the time of this writing, additional tests were in the process of approval in the United States. Most of these assays have been used extensively in Europe, but the USDA had not comprehensively adopted these assays before the 2003 occurrence of BSE in North America. However, a gap in all testing procedures is the reliance on a sample of brain taken from an older, dead animal. Since infection of cattle occurs in young calves, the ability to sensitively detect young infected calves using a sample from a live animal would be a significant advance. Current research on diagnostics focuses on the development of a sensitive live animal test. Unfortunately, this is a challenge in BSE, since experimental data suggest that the abnormal BSE prion is either not present or is rarely and inconsistently present in blood and lymphoid tissue, unlike scrapie which is present in both blood and lymphoid tissue or chronic wasting disease (CWD), which is consistently present in lymphoid tissue (Hunter et al., 2002; Hibler et al., 2003). For this reason, current research focuses on finding surrogate markers of BSE infection by understanding the host response that leads to conversion of the normal prion to the abnormal variant.

PREVIOUSLY UNKNOWN AGENTS

Severe Acute Respiratory Syndrome (SARS) Coronavirus

Severe acute respiratory syndrome (SARS) is a viral respiratory illness caused by the SARS coronavirus (SARS CoV). SARS was first reported in Asia in February 2003. Over the next few months, the illness spread to more than two dozen countries in North America, South America, Europe, and Asia before the SARS global outbreak was contained. (See Box 3-5 for a description of the outbreak.) According to the World Health Organization (WHO), a total of 8,098 people worldwide became sick with SARS during the 2003 outbreak. Of these, 774 died. In the United States, only eight people had laboratory evidence of SARS-CoV infection. However, in addition to the direct health costs of treating those people and testing others, as with FMDV and other exotic diseases, SARS had devastating effects on both global travel and trade, and its social and economic global impacts were disproportionate to the number of actual fatalities. Due to concern about spread of the disease, there were increased medical facility costs to prevent the spread of the highly contagious disease, altered travel plans (both business and pleasure), and additional precautions taken in airline travel. The infection of large numbers of health care workers, coupled with exhausting demands on the remaining staff, created additional burdens for the severely stressed health care systems. Economists have estimated the global economic loss from SARS at close to $40 billion in 2003 (Lee and McKibbin, 2004).

International Collaboration, Coordination, and Outreach

Lacking the research and investigative capacity to control the SARS epidemic, WHO elicited public health service partners from countries such as the United States, the United Kingdom, Germany, and France. WHO's international Global Alert and Response Network (GOARN) is a virtual network of 11 leading, well-equipped, and high biosecurity infectious disease labs in nine countries established primarily to address influenza outbreaks. GOARN was instrumental in spearheading laboratory efforts: these labs were connected by secure web sites and daily teleconferences to identify the causative agent of SARS, develop diagnostic tests, and collect and analyze clinical and epidemiological data on SARS. The U.S. CDC established other virtual teams in the United States, eliciting advice from medical experts, epidemiologists, and virologists, including both biomedical and veterinary coronavirus experts. Highly trained personnel from the CDC were dispatched to outbreak areas to assist in infection control, and numerous CDC employees were involved in all aspects of the response to SARS. To educate the public by countering rumors with reliable information, both WHO and the CDC provided factual information on SARS through updated web sites, satellite broadcasts, frequent presentations to the news media, and public response hotlines for clinicians and the general public.

These exceptional international laboratory efforts led to the rapid identification of a new coronavirus as the causative agent of SARS by April 16, 2003, only about 1 month after the initial WHO global alert (Drosten et al., 2003; Ksiazek et al., 2003; Peiris et al., 2003; Poutanen et al., 2003). Although vaccines or antivirals to prevent or control SARS infections were lacking, the SARS epidemic was countered by classical infection control and containment methods. These included screening of individuals for symptoms (fevers) with isolation, quarantine, and effective clinical management of symptomatic patients, followed by contact tracing and 10-day quarantine of known contacts. Implementation of effective surveillance measures, identification of the causative agent of SARS as a coronavirus, and containment of the SARS epidemic were attributed to this unparalleled level of global cooperation. Two features of the global SARS outbreaks include the following:

- SARS had devastating social and economic global impacts disproportionate to the number of actual fatalities and affecting both global travel and trade.
- International collaboration and communication among agencies and with scientists in established laboratory networks with prior working relationships and access to state-of-the-art equipment and the required biosecurity level were key elements for rapid and successful SARS diagnosis and control.

BOX 3-5
The 2003 SARS Outbreak

Atypical pneumonia cases, later characterized as SARS, first occurred in the Guangdong Province of China in mid-November 2002. Early data suggested a possible zoonosis, with the earliest SARS cases detected among workers dealing with exotic food animals. In subsequent studies, SARS-like coronaviruses were detected or isolated from two wild animal species in live animal markets, the masked palm civets (*Paguma larvata*) and raccoon dogs (*Nyctereutes procyonoides*), although the natural animal reservoir remains uncertain. Further studies have documented that nonhuman primates, ferrets, domestic cats, mice, and hamsters are also susceptible to experimental SARS coronavirus (Fouchier et al., 2003; Guan et al., 2003; Martina et al., 2003; Glass et al., 2004; McAuliffe et al., 2004; Roberts et al., 2005); pigs in China were naturally infected with a human-like SARS strain lacking the 29-nt insertion (Chen et al., 2005). Conversely, civets are susceptible to infection and disease with at least two human strains of SARS CoVs: one early human isolate (GZ01) with the 29-nt insertion like the civet strains, and one later human strain (BJ01) without the 29-nt insertion (Wu et al., 2005). Similar exotic animal markets also provided a breeding ground for recent influenza outbreaks in Hong Kong. The unsanitary conditions in live animal markets in China (and elsewhere) foster an environment conducive to the emergence of new zoonotic and animal diseases and likely played a role in SARS transmission from animals to humans (Peiris et al., 2004; Xu et al., 2004).

Global spread of the SARS epidemic was triggered on February 21, 2003, by a superspreading event in the Metropole Hotel in Hong Kong by an infected physician from Zhongshan University in China. Within 24 hours, he infected others at the hotel, who then carried SARS to Singapore, Vietnam, Canada, Ireland, and the United States, besides elsewhere in Hong Kong. Based on WHO estimates, this superspreader initiated a chain of infection involving nearly half of the 8,000 cases in more than 30 countries. On March 12, 2003, the World Health Organization (WHO) issued a global alert describing atypical pneumonia cases (severe acute respiratory syndrome or SARS) in Hong Kong and Vietnam and initiated worldwide surveillance. In an unprecedented move on March 15, WHO issued a travel advisory regarding high-risk areas where SARS outbreaks had been detected. The agency continued to issue travel advisories and advise airline passenger screening from high-risk areas through mid-April 2003.

From mid-March to April 2003, a second major outbreak of SARS occurred in another location in Hong Kong, the Amoy Gardens Apartments, with 321 people ultimately infected (Chim et al., 2003). This outbreak was more severe clinically, with more diarrhea cases (73 percent), higher intensive care unit admissions (32 percent), and higher mortality rates (13 percent) than in the Metropole Hotel outbreak. Environmental factors (in a faulty sewage system) were postulated to have contributed to the virus spread in the Amoy Gardens via aerosolized fecal material.

SARS did not spare developed countries, even ones with modern public health systems and significant resources. Canada had an outbreak of SARS on February 23, 2003, imported from Hong Kong. A second outbreak followed in mid-May 2003 after a lapse in infection control. Also because of a delayed initial response, SARS was not controlled in China until late June 2003. By that time, over 5,000 cases had been reported. On July 5, 2003, after control of a Taiwan outbreak, WHO reported that the global SARS epidemic had been contained. However on September 8, 2003, a single case of SARS was reported in Singapore (confirmed by the CDC). This individual likely became infected via laboratory-acquired exposure to SARS coronavirus, illustrating the need for strict adherence to laboratory safety procedures required for work with BSL-3 level pathogens. The widespread distribution of SARS coronavirus samples in international labs highlights the need for vigilance in the inventory of these virus stocks. Also, adequate laboratory supervision and facilities are required to avoid future laboratory acquired infections as a possible source of new SARS outbreaks.

In December 2003–January 2004, several new SARS cases reemerged in Guangdong Province, China (Normile, 2004). For at least one case, no risk factor was identified such as a link to civets. Other postulated reservoirs including rats and cats were tested, but no final conclusions were drawn concerning the origin of this reemergent case. However, sequence data suggested that the reemerged SARS coronavirus strains were more like the civet isolates (Normile, 2004), and China ordered the destruction of large numbers of civets in its wildlife markets (Watts, 2004). Recent data based on serology suggest that some SARS antibody seropositives occurred in Hong Kong in 2001 before the documented SARS outbreaks, suggesting that low numbers of subclinical SARS infections likely occur (Zheng et al., 2004). Thus both animal reservoirs and subclinically infected humans remain potential sources for the reemergence of SARS.

Links between Veterinary Research and Public Health

The veterinary coronavirus (CoV) research community provided important resources and an extensive background perspective on coronavirus biology, contributing to an improved understanding of SARS evolution and infections. From the isolation of genetically similar CoV from civets and raccoon dogs in live animal markets in China, scientists postulate that SARS evolved from a wild animal host (Guan et al., 2003). Many of the earliest SARS cases and SARS antibody detections occurred among workers who had contact with exotic food animals in Guangdong animal markets (Guan et al., 2003; Xu et al., 2004). The unsanitary and crowded conditions, with multiple species in close contact in live animal markets in China (and elsewhere), fosters an environment conducive to the emergence of new zoonotic and animal diseases and likely played a role in the putative transmission of SARS from animals to humans. The published evidence and epidemiological data suggest that SARS was a probable zoonosis (Guan et al., 2003; Peiris et al., 2004; Xu et al., 2004; Wu et al., 2005). These markets also provided a breeding ground for the 1997 avian influenza outbreaks in humans in Hong Kong (Hampton, 2004). Prior studies of animal CoVs have shown that interspecies transmission of CoVs is not unprecedented (Tsunemitsu et al., 1995; Ismail et al., 2001). However, the determinants of CoV host-range specificity and the potential of wildlife as reservoirs for emergence of other CoV strains of potential threat to public or animal health are unknown.

In addition, respiratory and enteric CoV infections in the natural animal host (swine, cattle, poultry) have provided important information on CoV disease pathogenesis and possible potentiators for increased disease severity applicable to SARS CoV infections. Enteric CoV infections alone frequently cause fatal infections in young animals. However in adults, respiratory CoV infections are more severe or often fatal when they are combined with other factors including stress and transport of animals (shipping fever of cattle), high exposure doses, aerosols, treatment with corticosteroids, and other respiratory co-infections (viruses, bacteria, bacterial lipopolysaccharides) (Saif, 2004). Such variables may influence the severity of SARS or contribute to the phenomena of superspreaders.

Thus coronaviruses, largely ignored by the biomedical research community and public health funding agencies because of low-impact human infections, are a recognized and significant cause of potentially fatal respiratory and intestinal infections in animals. This knowledge base of animal coronavirus pathogenesis, vaccines, and basic studies of coronavirus replication strategies and development of infectious clones contributed to rapid progress in characterization of SARS coronavirus and is critical for future SARS vaccine and antiviral strategies. Lessons of the global SARS outbreaks include the following:

- Wild animals transported from their native habitat or forms and introduced into live animal markets may harbor unknown disease agents such as SARS that are transmissible to humans.
- Veterinary science research contributed to the understanding of SARS CoV pathogenesis, particularly its potential for interspecies transmission and fatal disease, which had significant implications for public health.
- There is a need for greater collaboration between the animal health and medical research communities in studying the pathogenesis of previously unknown and likely zoonotic agents such as SARS. In the case of SARS, research ties and interagency funding and cooperation are lacking to promote collaborative infectious and zoonotic disease research between biomedical and veterinary scientists and to provide trained biomedical and veterinary public health personnel. Furthermore, funds are lacking to study disease pathogenesis in the appropriate animal host and to investigate zoonotic diseases including identifying animal reservoirs and the mechanisms and chain of interspecies transmission.

Diagnostic Techniques

Classical virological techniques (such as electron microscopy [EM], cell culture, and immunofluorescence), as well as new molecular approaches (microarray) were essential for identification of the previously unrecognized SARS coronavirus. Many diagnostic labs are phasing out EM and/or cell culture facilities because of costs or lack of trained personnel and putting more emphasis on molecular techniques. However, most current molecular approaches (RT-PCR with specific primers) or serology are designed to detect known, but not unknown or new, pathogens. In addition, the focus of public health agencies (CDC, WHO) is on development and validation of diagnostic tests for SARS applicable to humans. These tests, especially antibody assays, lack validation for use in various animal species or lack animal coronavirus controls, creating difficulties for assay development and interpretation of results from testing wild or domestic animals. Little funding is available from federal agencies to stimulate development of new or improved diagnostic assays for humans and animals, including ones targeted to identification of microbial nucleic acid signatures, or to study the relationships of pathogens common to animals and humans or their disease mechanisms or persistence in the animal host. Development and availability of standardized validated test protocols, reagents, and controls is essential for reliable diagnostic tests to monitor SARS cases in the early stages of infection. Accurate tests also require reagents and testing of both human and animal CoV strains as controls to eliminate confounding cross-reactivities and to identify the animal reservoir for SARS.

ENDEMIC DISEASES: AVIAN INFLUENZA, CHRONIC WASTING DISEASE, AND WEST NILE VIRUS

A large array of indigenous diseases could serve as case studies in the category of endemic disease. The committee selected three diseases, including avian influenza (AI), which it chose for four reasons: (1) it has been a recurring disease problem that has caused significant economic loss to the poultry industry in the United States; (2) strains of the influenza virus that could cause a human pandemic can emerge; (3) it occurs in many countries; and (4) strains are harbored in migratory waterfowl. The committee looked at chronic wasting disease (CWD) to illustrate the potential of wildlife disease to decimate wildlife populations, the danger of translocation of animals in spreading diseases, and public concern about transmissible spongiform encephalopathies (TSEs). CWD was selected as an example of a serious disease that is, at this point, exclusively a problem of wildlife and game farms, specifically elk and deer (cervids) in North America. Diseases of wildlife are a concern, not only because wild animal populations can be depleted by diseases (with potential ramifications for biodiversity and ecological integrity), but also because wild animals can be reservoirs of indigenous and exotic infectious diseases of domestic animals and reservoirs of human zoonoses. Game animals can be a source of infectious disease agents into the human food chain. The third disease selected for analysis, West Nile virus (WNV), is an example of a disease that will likely persist in North American wildlife and remain a constant threat to domestic animals and people. All of these diseases illustrate the consequences of increasing biological connectedness in today's world. At one time these agents may have been considered as newly emergent, but they have now become firmly established in North America and for the purposes of this report are considered endemic.

Avian Influenza

Of the diseases examined in this chapter, AI (also called "bird-flu") could arguably be the most representative for evaluating comprehensiveness of the animal health infrastructure. The virus has the potential to impact public health, production animals (e.g., poultry and swine industries), and wildlife (e.g., waterfowl and migratory birds). It is on the list of potential biothreat agents, and the CDC and state departments of public health address it annually in their Pandemic Disease Plans. Enhanced government awareness of biological threats and the associated "culture of fear" following the September 11, 2001, attacks and the terrorist anthrax letters is still very apparent in the public media, yet the most probable and real threat, that of a pandemic due to influenza virus, has received relatively little attention beyond the scientific community. Since the 1997 H5N1 avian influenza virus adapted to humans, the academic and public health communities have warned of the potential for an influenza pandemic, a global outbreak that could mimic the 1918 Spanish flu outbreak. The 1918 Spanish flu caused the single deadliest epidemic in history, killing between 20 and 50 million people in just half a year (IOM, 2003, 2005). In 2004, a virulent strain of H5N1 avian influenza virus spread through numerous Asian countries. The virus jumped species 6 months into the outbreak, making humans susceptible to infection (CDC, 2004c). By January 2005, the situation in Southeast Asia had worsened as the H5N1 virus continued to spread into the human and bird populations. As of January 2005, the H5N1 avian influenza virus killed 34 of 47 infected humans and resulted in the death and depopulation of over 100 million birds, primarily commercial poultry, as well as uncounted numbers of wild birds. More significantly, in September 2004, AI apparently spread not from a bird to a human, but directly from an infected child to her mother (Ungchusak et al., 2005). The demonstrated ability of the virus to spread from one human to another makes the possibility of a pandemic a significant threat.

Influenza prevention measures for susceptible animals, primarily swine and poultry, rely on vaccination, quarantine, and depopulation. These measures failed to contain the 2004 spread of Asian H5N1 influenza. In the public health domain, an October 2004 announcement that 48 million of the expected 250 million global doses of influenza vaccine would not be available due to closure of a vaccine producer's facility clearly amplifies the threat of unchecked spread should the virus establish itself in humans (HHS, 2004).

Nature of the Pathogen

Avian influenza viruses are endemic in wild bird and migratory waterfowl populations and can be transmitted to domestic poultry. Table 3-2 provides a summary of key events linked to the identification and spread of the disease. Some influenza subtypes are also capable of infecting mammalian species, in particular swine and humans. The pathogenicity of these viruses (of which there are 15 known subtypes identified as H1–H15) can be classified as low to high, dependent on the severity of disease caused. Of the 15 avian influenza virus subtypes, H5N1 is of particular

TABLE 3-2 Timeline of Key Influenza Events

1878	• Fowl plague (FP) was described as a serious disease in chickens in Italy.
1918	• Spanish flu (influenza A) pandemic claimed 20 to 50 million lives worldwide in less than a year and ranks among the worst disasters in human history. In the United States alone, an estimated 1 in 4 people became ill and 675,000 people died (Crosby, 1989). Recent studies now suggest this historic pandemic was associated with interspecies transmission of an avian influenza virus (Hampton, 2004; Stevens et al., 2004).
1955	• Fowl plague virus was determined to be one of the influenza viruses.
1984-1985	• Outbreak of avian influenza virus H5N2 in poultry in the Northeast United States. It initially caused low mortality, but within 6 months had mutated to a highly pathogenic virus causing nearly 90 percent mortality. The outbreak cost over $65 million and resulted in the destruction of 17 million birds.
1992	• An avian influenza virus H5N2, identified as "low pathogenecity" in Mexico, mutated to a highly pathogenic form and continued to spread until 1995. In 1999, an Italian H7N1 virus had a similar pattern of mutation over a 9-month period and was not controlled until 2001. The 2001 Italian losses are estimated at 13 million birds.
1997	• The first documented AI infection of humans occurred in Hong Kong when the H5N1 strain caused severe respiratory disease in 18 humans, of whom 6 died. Extensive investigation determined that close contact with live infected poultry was the source of the human infection. Studies at the genetic level further determined that the virus had jumped directly from birds to humans but had only very limited human-to-human spread.

concern because it mutates rapidly and has a documented propensity to acquire genes from viruses infecting other animal species (WHO, 2004). The majority of avian influenza viruses have low pathogenicity, typically causing little or no clinical disease in infected birds, particularly migratory waterfowl, which serve as a reservoir of the virus. The highly pathogenic strains may be associated with mortality close to 100 percent (Easterday et al., 1997; WHO, 2004). The highly pathogenic influenza virus subtypes can cause significant economic losses to poultry, impinge on international trade, and, if transmitted to humans, pose public health risks with the potential to initiate deadly human influenza pandemics. The virus is additionally considered a potential biothreat agent based on its abil-

2003	• A 57-year-old veterinarian who visited a poultry farm affected by the H7N7 strain died on April 17, 2003, of acute respiratory distress syndrome in the Netherlands. H7N7 influenza virus was isolated from the patient. No other respiratory pathogen was detected in a series of laboratory tests (WHO, 2003). • West Virginia initiated preemptive disease eradication efforts in cooperation with the USDA after observing mutation of a circulating low pathogenecity H7N2 virus toward high pathogenecity. Texas initiated similar control efforts directed at low pathogenic avian influenza a few years earlier.
2004	• In early 2004, a similar situation played out in the Northeast United States associated with a rapidly spreading, low pathogenecity H7N2 influenza variant. More than 400,000 chickens in two states were destroyed, and additional farms quarantined following detection of the H7N2 influenza virus. International trade restrictions were also imposed on the United States within days of the first cases being reported. • In Canada, at least 17 million poultry died or were destroyed to contain the spread of Canada's first reported detection of highly pathogenic influenza, an H7N3 virus, which also occurred as a low pathogenic form of the virus on the same farm. • In Texas, 8,900 chickens were destroyed following detection of a highly pathogenic variant of H5N2, which was the first detection of a highly pathogenic strain since 1984 in the United States. • Thijs Kuiken and colleagues at the Erasmus Medical Centre in the Netherlands found that cats could become infected and spread the avian influenza virus H5N1 (Kuiken et al., 2004). • Beginning in January 2004 and continuing into 2005, an outbreak of highly pathogenic H5N1 avian influenza spread through 11 Southeast Asian countries, affecting millions of birds, including multiple avian species. By January 2005, the outbreak had resulted in the deaths of 34 of 47 people infected in two different countries (WHO, 2005).

ity to spread easily and rapidly as a respiratory infection in both animal and human populations. Influenza virus outbreaks can directly involve federal, state, and local agencies as diverse as those that deal with homeland security, public health, agriculture, interior, commerce and trade, natural resources, and environmental quality.

Interspecies Transmission (Particularly to Humans)

In 1997, the first documented case of avian influenza transmission to humans occurred in Hong Kong, affecting 18 people, killing 6, and resulting in the destruction of 1.5 million birds in efforts to eliminate the variant virus. Since 1997 several instances of bird-to-human transmission have occurred, and recent studies suggest that the historic Spanish flu pandemic of 1918 was associated with interspecies transmission of an avian influenza virus (Hampton, 2004). In 1999 (Hong Kong and China H9N2), 2003 (Hong Kong H5N1, Netherlands H7N7, Hong Kong H9N2), and 2004 (Viet Nam and Thailand H5N1), avian influenza was transmitted from birds to humans, in some cases resulting in death of infected individuals. In the recent bird-to-human scenarios, little transmission of the avian influenza virus from human to human was detected. However, public heath officials fear the possibility of a human influenza virus and an avian influenza virus infecting the same individual and "reassorting" or trading genes, to produce a highly lethal virus capable of rapidly spreading in the human population.

Cause for Concern—Possibility for Human-to-Human Transmission and Mutations of Low-Pathogenic Viruses

Should an avian influenza virus unfamiliar to the human population gain the ability for human-to-human transmission, the predicted outcome is a pandemic with serious disease and death globally. In the three influenza pandemics that occurred during the past 100 years, all spread worldwide within 1 year, causing significant social and economic disruption. (The 1918 Spanish flu resulted in nearly 675,000 deaths in the United States, the 1957 Asian flu caused 70,000 deaths, and the 1968 Hong Kong flu caused 34,000 deaths.) Once established in the human population, influenza viruses tend to persist and significantly affect human health for years. The CDC and WHO continuously monitor and react to influenza viruses, with the primary goal to watch spread and determine vaccine strains, but also with the goal of preventing or controlling the emergence of a new and potentially pandemic influenza virus.

As the 1984–1985, 1992, and 2001 outbreaks in poultry illustrate, influenza viruses of low pathogenicity have the capacity to mutate into highly pathogenic strains, sometimes after very short periods of circulation in poultry populations. Aggressive surveillance, detection, and disease control, generally including total depopulation of poultry in the area, are considered critical to minimize transmission, control economic losses, and eliminate the public health risks associated with human exposure to highly pathogenic avian influenza viruses. As Table 3-2 indicates, individual states have initiated preemptive disease eradication efforts in cooperation with USDA APHIS and their poultry industries after observing mutation of a circulating low pathogenecity H7N2 virus toward high pathogenecity. Though effective, the low pathogenecity influenza eradication programs were initially hindered by lack of defined USDA regulatory authority for endemic diseases and absence of an established indemnity program for the depopulated birds.

In February 2004, prompted by the H5 avian influenza outbreaks in Southeast Asia, HHS and USDA officials cooperatively announced a ban on importation of birds from eight Asian countries. Globally, public health as well as animal health agencies have closely followed the appearance and movement in 2004 of H5 avian influenza viruses in poultry and are particularly concerned about its rapid spread through Asia, where acceptance or compliance with slaughter-based control efforts is considered not economically feasible or socially acceptable. The millions of affected birds, commingling of different avian and mammalian species, difficulty in protecting poultry workers from respiratory exposure, and the recognition of bird-to-human transmission have placed the global community on high alert for the potential evolution of a pandemic influenza virus. The committee drew the following conclusions regarding AI:

- Gaps in scientific knowledge, focused and applied research, understanding of disease risks, and lack of access to validated rapid detection methodologies have complicated and jeopardized effective and timely responses to AI.
- Lack of a standardized reporting mechanism among animal health agencies has delayed laboratory findings and epidemiological investigations.
- Though clearly defined for highly pathogenic avian influenza, the regulatory lines of authority are not defined for endemic (low pathogenecity) avian influenza, hindering the nation's ability to prevent a potentially devastating disease situation.

Chronic Wasting Disease

As a prion-associated transmissible spongiform encephalopathy (TSE), CWD belongs to a group of diseases that merit careful study and monitoring (NRC, 2004a). Over a relatively short time span, CWD has become a major problem in some U.S. western and midwestern states, affecting both farmed and wild cervids, thus impacting the markets for farmed cervids and cervid products and for wildlife-related recreation and aesthetics. While there is no conclusive evidence to date that a CWD prion has caused naturally occurring disease in domestic animals or people, research needs to continue to determine whether this disease is a threat to other than cervid species (Belay et al., 2004).

Occurrence and Transmission

CWD was first identified as a disease syndrome in 1967 in a mule deer in a research facility in Colorado that had been populated by wild deer captured in that state (Williams and Young, 1980). It is likely the disease existed in nature but went unrecognized. About 10 years later CWD was determined to be a TSE (Williams and Young, 1980; Williams et al., 2002; Belay et al., 2004).

CWD occurs in free-ranging native elk, mule deer, and white-tailed deer and was first reported in northeast Colorado and southeast Wyoming in the 1980s. Its prevalence in this area over the period 1996–1999 was found to be about 5 percent in mule deer, 2 percent in white-tailed deer, and less than 1 percent in elk (Miller et al., 2000). Since 2002, the wider application of testing of hunter-killed or other animals has uncovered endemic loci of infected animals in western Colorado, Nebraska, New Mexico, South Dakota, Utah, Wyoming, Wisconsin, Illinois, and New York, and in Saskatchewan, Canada. The prevalence of CWD in endemic areas has been estimated to be less than 1 percent for elk and to vary from less than 1 percent to 15 percent for mule deer (Williams et al., 2002).

CWD has been a major problem for the cervid farming industry. The first diagnosis on an elk farm occurred in South Dakota in 1997 and marked the beginning of more widespread testing (USDA APHIS-VS, 2005). Subsequently the disease has been found on a substantial number of game farms in other parts of the United States (Colorado, Kansas, Minnesota, Montana, Nebraska, New York, Oklahoma, and Wisconsin) and Canada (Alberta and Saskatchewan) (AVMA, 2005; Belay et al., 2004). It is reasonable to assume that CWD has been spread in large part through trade associated with the growth and development of the game farming industry. Spread from game farms to free-living cervids appears to have occurred in at least some but not all situations where game farms have been established. The widespread dissemination of CWD among game farms in Canada resulting from importation of infected elk is beyond dispute (Bollinger et al., 2004). However it occurred, CWD has emerged as an immediate and serious threat to wildlife resources that generally are highly prized by society and serve as the basis for substantial economic activity.

Pathogenesis

CWD is associated with the accumulation of abnormal prions in brain and lymphatic tissue. While not yet universally accepted, the "infectious" agent is believed to be an abnormal prion (Belay et al., 2004). The disease has a long incubation period, probably 15 months or more before the appearance of clinical signs (Belay et al., 2004). Affected animals lose weight and may show neurological signs. Definitive diagnosis is based on direct laboratory examination of brain or lymphatic tissue by immunodiagnostic techniques that can determine the presence of the agent in clinically normal animals as well as those showing signs. These tests are the basis of epidemiological surveys for prevalence of the disease in culled and hunter-killed animals.

It is likely that CWD is transmitted among cervids by urine, feces, and saliva (Miller et al., 2004). The CWD agent remains infective in contaminated premises for at least 2 years and presumably this would also be true in the natural environment (Miller et al., 2004). No evidence suggests that other species can acquire CWD under natural circumstances or when housed in direct contact with affected deer or on contaminated premises. There is no known risk to humans from consuming meat from deer or elk, but prudence makes consumption of meat or nervous tissue from animals known to be infected unwise until more is known about this disease (Belay et al., 2004).

APHIS has developed procedures to validate and approve testing procedures for diagnosing CWD. There is a danger of moving animals to new habitats in the absence of the means to certify their health with a high level of assurance or guarantee of their quarantine in the new location. The history of CWD illustrates that controlling the translocation of captive wildlife is of paramount importance in preventing the spread of wildlife disease.

The national response to CWD has been guided since 2002 by a Plan for Assisting State and Federal Agencies and Tribes in Managing CWD in Wild and Captive Cervids (USDA APHIS, 2002), developed by a task force of federal and state wildlife management agencies. Subsequently a group representing a broad constituency of universities, professional organizations, and interest groups developed action items. The task force has directed attention to six broad areas of activity: communications, dissemination of technical information (including education), diagnostic methodology, disease management, research, and surveillance. Both USDA and DOI have provided funding for CWD programs (FYs 2003–2005). Implementation of the plan has been coordinated through the Multi-States CWD Working Group, U.S. Animal Health Association, and the International Association of Fish and Wildlife Agencies. The DOI (NPS, USGS, FWS) and USDA (APHIS, CSREES) have established interagency work groups to deal with specific issues. Interested nongovernment organizations have banded together to form the CWD Alliance to keep stakeholders appraised of developments. The formation of the CWD Alliance by conservation organizations and the development of its web site (www.cwd-info.org) have been particularly effective in providing a source for accurate information among nonexperts. Also, a substantial number of continuing education programs for biologists and others working on CWD have been developed or are in preparation by agencies involved in the Plan.

Diagnosis

Immunohistochemistry (IHC) tests of tissue samples remain the approved means for definitive diagnosis of CWD. ELISA based methods have been developed for screening samples more rapidly, but positives must be verified by IHC. Research continues on developing methods that could be used to diagnose infection in live animals or be more cost-effective. By April 2003, the NVSL had approved 26 university or state diagnostic laboratories to provide CWD testing, a number that accommodated demand prior to the onset of BSE surveillance. Currently the system is stretched to or somewhat beyond capacity, and additional laboratories may be approved.

Prevention

Steps are being taken to prevent or limit the spread of CWD. All 50 states and several federal and tribal agencies have conducted monitoring and management activities partially funded by APHIS through cooperative agreements. As of June 2004, 24 state wildlife management agencies adopted a policy set out in *Multi-state Guidelines for Chronic Wasting Disease Management in Free-ranging White-tailed Deer, Mule Deer and Elk*. The APHIS-proposed rule, *Chronic Wasting Disease (CWD) Herd Certification Program and Interstate Movement of Captive Deer and Elk,* is pending. Some states have put restrictions on baiting and feeding of free-ranging cervids.

Experience in both the United States and Canada indicates that eradication of CWD in *farmed* cervids should be possible (Bollinger et al., 2004). On the other hand, containing CWD in *free-living* cervids and preventing its spread to contiguous regions free of the disease will be extraordinarily difficult (Gross and Miller, 2001). Of paramount importance is controlling the translocation of animals or infected material. It has been recognized this will require not only appropriate regulations, but also an educated public that will not be tempted to ignore restrictions on the movement of cervids.

Containing CWD will require a well-coordinated management program supported by an aggressive research program that defines the population density, buffer zone size, and time required to render contaminated environments free of infectivity to arrest spread to contiguous areas. It has been suggested that deer densities of less than 1 square kilometer of critical habitat will likely have to be maintained for prolonged periods of up to 10 years or more to create such buffer zones and prevent spread from endemic areas (Bollinger et al., 2004). Through an examination of CWD, the committee learned that:

- The infrastructure to detect, diagnose, and prevent wildlife diseases is an essential element in the nation's framework for dealing with animal diseases and its consequences. In the past, it has been flawed by inadequate (1) coordination of relevant agencies, (2) diagnostic expertise, (3) research support, and (4) education and training of professionals. Experience with CWD, and other diseases of wildlife such as WNV or the pathogens they may harbor, such as avian influenza, has highlighted these deficiencies and provided the motivation to correct these inadequacies.
- In purely biomedical terms, the discovery of CWD has provided a research opportunity in comparative medicine for gaining greater insight into the pathogenesis of TSEs. The recent creation by the White House of an interagency working group on disease-causing prions to identify gaps in knowledge is timely.
- The spread of CWD, taken together with the history of the spread of raccoon rabies and of monkeypox, provides compelling contemporary evidence of the need for much more effective control of translocation of wild animals in preventing the spread of animal disease. It reaffirms an essential strategy for disease control that has been recognized since the establishment of the first U.S. national animal health agency, the Bureau of Animal Industry, in 1884. The translocation of wildlife, both indigenous and exotic, is fraught with every bit as much risk as translocation of domestic animals, especially since organisms that are symbiotic in wildlife may be pathogenic in people and/or domestic animals.

West Nile Virus

The introduction and rapid spread of WNV (caused by a single stranded RNA flavivirus) across the North American continent is a simple and powerful illustration of the growing importance of a zoonotic disease that is harbored and spread in wildlife in an increasingly interconnected world. The disease was first described in Uganda in 1937. Since that time it has become endemic in other parts of Africa, Southwest Asia, and Europe, where it appears to cause relatively few cases of disease in people, domestic animals (horses and geese), or wildlife.

WNV appeared in New York in 1999 from an unknown source. By 2003 it had spread across the continent in a naive bird population that sustained the infection and allowed it to be spread by mosquitoes that feed largely on birds. While WNV primarily infects birds, and while most species are apparently resistant, corvids are particularly susceptible and many died from the infection. Surveillance for the virus in dead crows proved to be a better indicator of looming human and animal infection than isolation of the virus from trapped mosquitoes or sentinel flocks of chickens. Humans and horses become infected when species of mosquitoes that feed on both mammals and birds (e.g., some *Culex spp*) are favored by climatic circumstances. About 80 percent of the people who become infected with WNV will have no symptoms and the vast majority of the remaining 20 percent will have mild disease symptoms (CDC, 2004e). Unfortunately, a few acquire encephalitis that can be fatal. In 2002 and 2003, WNV infected 15,300 and 5,200 horses in North America, respectively (USDA APHIS, 2004b). It can be expected that the disease will become entrenched permanently in some North American ecosystems. The industrial and research establishment in the United States responded to this outbreak by developing an effective vaccine to protect horses and a rapid screening test that provided for efficient and cost-effective presumptive diagnosis. The committee found that globalization brings increasing risk from diseases such as WNV that have hitherto not been present in the United States but are transmissible to people and domestic animals by wildlife. It is unlikely that all such diseases can be prevented from entering and staying in the country.

North America has large, pristine, nonimmune populations of animal hosts, humans, multiple efficient and competent vectors, and a favorable environment for disease transmission and spread. The introduction of WNV became the key catalyst and final element needed to produce one of this country's most significant epidemics. Every year since 1999 has led to new information about this epidemic.

West Nile virus has an extremely broad host range replicating in birds, reptiles, amphibians, mammals, and numerous mosquito vectors. While WNV is primarily a vector-borne zoonotic disease that is maintained through a bird-to-bird transmission cycle via mosquito vectors, new information has also revealed that this flavivirus can be transmitted by other means: for example, contaminated blood products, organ transplantation, maternal transmission via breast milk and intrauterine, percutaneous exposure in a lab setting, and, at least experimentally, direct horizontal transmission between and among birds due to exposures via fecal shedding.

The WNV epidemic is still evolving and not well understood. The recent recognition of a very large host range, numerous potential vectors, nonvector modes of transmission, and the potential movement into Central and South America via bird migrations has revealed that WNV is a much greater threat and more difficult to control than initially realized. There is every indication that WNV is becoming established as an endemic zoonosis and will be a permanent part of the animal and human medical landscapes in the United States and beyond for many years to come.

The scope, scale, intensity, and consequences of this ongoing disease problem are unprecedented for an arbovirus (an arthropod-spread virus). WNV is a remarkable example of an emerging zoonosis that involves the dynamic and complex interface among domestic animals, humans, and wildlife species. The United States was an ideal setting for this epidemic, and the virus continues to adapt and move into new ecological settings that produce new selective pressures and exposures to new hosts and vectors that constantly change the nature of this disease. The final chapter on this disease has not been written because the story continues to unfold and will continue to do so in unprecedented ways.

Important lessons to be learned from the WNV epidemic include the following:

- With the propitious epidemiological conditions for arboviruses in the United States, epidemic events can result in long-term endemic disease problems in multiple sites and species. The ease of spread of WNV also provides a living model and example of what a purposeful introduction of a vector-borne zoonotic disease might do. Rift Valley fever, for example, has epidemiological similarities to WNV that should be of serious concern.
- There is a need for a better understanding of most aspects of WNV, including its adaptations, environmental survival, host and vector range, and nonvector transmission modes. Ecology, wildlife dynamics, and epidemiology are among the scientific disciplines that need to be addressed by the animal health framework in the future.
- WNV offers further evidence that the veterinary profession and animal health organizations must develop the expertise, knowledge, and skills needed to address the implications of zoonotic diseases.
- The epidemic illustrates the need for strategic and collaborative partnerships between government agencies, and especially among animal and human health officials and communities. Those partnerships also need to extend globally.

The confluences of human and animal health, along with wildlife, create new opportunities for pathogens to emerge and reemerge. Microbes will adapt, gain competitive ecological advantages, and threaten populations in novel and dangerous ways. WNV is a wakeup call to both human and animal health officials and organizations—it is a clarion suggesting that the past systems and operations in both of these communities will need to reconsider how and with whom they work.

NOVEL AND BIOENGINEERED PATHOGENS

The past two decades have witnessed the evolution and emergence of new strains or entirely new groups of animal or human pathogens. Many of these are thought to have emerged from interspecies transmission or as variants of established strains with new tissue tropisms. In this respect, emergent RNA viruses are notorious because of the presence of quasispecies or "swarms of virus within a viral population," high mutation rates due to lack of proofreading mechanisms for RNA polymerases, and the ability to generate genetic recombinant or reassortant viruses. Recent examples include the emergence of several RNA viruses in swine such as:

- The porcine respiratory and reproductive syndrome virus (PRRSV), an economically important arterivirus first isolated in 1991 in Europe and in 1992 in the United States, but with no known previous host and no prior serological evidence of previous infection detected in swine (Benfield et al., 1999).
- The porcine respiratory coronavirus (isolated in Europe in 1984 and the United States in 1989), an emergent virus that acquired a new tissue tropism (respiratory) and is a naturally occurring deletion mutant of TGEV, common, widespread swine intestinal coronavirus (Saif and Wesley, 1999).
- The porcine epidemic diarrhea virus (PEDV), a new group I coronavirus, which was reported initially in Europe in 1978 and in Asia in the 1990s as a major cause of diarrheal deaths in piglets, but still has not been detected in the United States (Pensaert, 1999).

These new viruses appear to have emerged independently in swine in Europe and the United States, suggesting that they have separate evolutionary origins. Equally troubling was the recent emergence in Malaysia of the new swine RNA virus, the Nipah virus, a zoonotic paramyxovirus that was transmitted to humans, necessitating the precautionary slaughter of large numbers of pigs (Chua, 2003). This latter virus was acquired by swine through interspecies infections, presumably from fruit bats. The identification of these novel porcine viruses required the use of classical (EM, cell culture) virological techniques, as well as molecular approaches, such as reverse transcriptase-polymerase chain reaction (RT-PCR).

If the challenge of diagnosing new, naturally occurring diseases were not already difficult enough, the prospect of identifying an intentionally-introduced recombinant (bioengineered) pathogen presents an even greater challenge. A bioengineered disease agent might, for example, be a chimera formed from unrelated pathogens that would confuse diagnostic tests by causing multiple reactivities. Specific antigen sites or genetic regions of a pathogen could be masked or deleted so that serological or genetic tests would result in false negatives. Therefore, a broad array of diagnostic tests with a wide spectrum of agent detection capabilities may be required to identify both novel pathogens, whether naturally occurring or intentionally engineered.

Identification of Novel Pathogens

Classical virological techniques and the fulfillment of Koch's postulates (for example, verifying a causative agent by using it to reproduce a similar disease syndrome in susceptible animal models) will remain important tools in disease diagnosis. However, rapid and unambiguous diagnostic tools are needed for early intervention in the event of a bioterrorism-related national crisis or for identifying novel pathogens. New molecular assays are evolving to meet those needs.

For example, the SARS coronavirus was identified using microarray technology that targeted consensus regions of gene expression within distinct virus families (Ksiazek et al., 2003). Pathogen specific detection methods using gene probes derived by genome mining, in conjunction with target or signal amplification, can yield methods sensitive enough to detect just a few organisms in suspect materials. The same technologies in combination with sequencing or microarrays can aid in genotyping pathogens, specifically in the diagnosis of antimicrobial agent resistance (Hamels et al., 2001). Of particular interest are microarrays designed to enable detection of major classes and families of organisms based on intergenic and polymorphic loci. Once such classification has been worked out, family-specific arrays may be used to scan for homologies and major differences to identify new pathogens. Another example is the use of mass spectroscopy and high-performance liquid chromatography methods on whole microorganism (bacteria, viruses, and protozoa) extracts to define protein profiles that can be matched to an existing database of fingerprints. This may enable classification of organisms into major families that can be used to further design molecular probes for a finer degree of discrimination (Bernardo et al., 2002; Huang et al., 2002; Frazier et al., 2003).

Rapid Monitoring of the Host Response to Pathogens

Serology (monitoring serum antibody responses to pathogens) is important in assessing seroresponders and the seroprevalence of infections to determine levels of exposure of the populations. However, it is inadequate for rapid disease diagnosis because antibody development usually takes 2–3 weeks.

Recognition of the rapid response (hours or days) in immunomodulatory factors such as cytokines in the blood to various pathogens suggests that microorganism-specific cytokine tests may permit more rapid monitoring of host-specific responses. An example is the adoption of assays to monitor for mycobacteria-specific antigen-induced secretion of the cytokine interferon-gamma in blood mononuclear cells as a more rapid diagnostic assay than slower cultivation protocols (weeks) required for isolation of *Mycobacterium bovis* from cattle (Stabel, 1996). Such tests could also provide rapid data on new immunosuppressive pathogens, naturally occurring or bioengineered, by revealing significant decreases in certain cytokine levels. Another example is the recognition that some immunomodulatory factors (the cytokine IL-4) coexpressed with a virus may increase the virulence of certain viruses (mousepox) in the host (Jackson et al., 2001). Monitoring of a dramatically up-regulated single cytokine level could signal the presence of such a bioengineered pathogen-cytokine recombinant or guide treatment with antibodies to the relevant cytokine or counter-regulatory cytokines. Besides of diagnostic value, innate immune factors (such as cytokines and interferons) occurring early (hours or first days) after infection, and preceding the acquired immune responses, might be manipulated as treatment modalities to reduce or prevent infection of contact or susceptible animals. Early innate immunomodulator intervention strategies might block or reduce pathogen infection and shedding, decreasing transmission to other animals or making contact animals more resistant to infection.

INTENTIONALLY INTRODUCED PATHOGENS AND DISEASES OF TOXICOLOGICAL ORIGIN

Of course, an act of bioterrorism need not involve a bioengineered pathogen. The intentional spread of known microorganisms or microbial toxins can be accomplished using the same routes as accidental introductions, which occur when disease agents are brought to new areas via the movement of air and water, fomites, vectors, infected animals, or animal products.

Currently, if naturally occurring, endemic agents were intentionally introduced into a new locale, investigatory agencies would be reliant solely on the pattern of outbreak to distinguish an attack, since there is currently no methodology to distinguish between intentionally versus accidentally introduced. The purposeful introduction of an exotic disease through channels of international commerce could also be disguised, since such an occurrence could as easily be the result of an accident as not. In the 2003 National Research Council report *Countering Agricultural Bioterrorism*, the Committee on Biological Threats to Agricultural Plants and Animals offered advice about animal diseases and animal disease vectors that could be used as agents of agricultural bioterrorism. Because the report discusses themes that are relevant to both intentionally introduced and naturally occurring disease and the potentially devastating consequences of a failure of the animal health network, we have included the conclusions of the report as Appendix D.

In addition to pathogens, toxins, chemicals, and radiological weapons might also be used to purposefully threaten animal health. These topics are outside the focus of this report. However, the committee acknowledges that toxic diseases (such as botulism and domoic acid) are an important concern for animal and human health. Diseases of toxicological origin present challenges different from those of other types of animal diseases. Domestic and wild animals are subject to toxins that occur naturally, for example in plants, or that are the result of human activity. Paralytic shellfish poisoning, domoic acid, *Pfeisteria* outbreaks, and related phenomena appear tied to coastal pollution and have the potential to make poisonous and inedible a growing proportion of the country's protein food supply. Therefore toxicology is an essential element in any program addressing animal disease.

SUMMARY

The lessons of past disease outbreaks and the prospects of future epidemics suggest that the animal health framework faces a formidable challenge in preventing, detecting, and diagnosing the spectrum of animal diseases, some of which have direct consequences for humans as well as animals. The challenge is multifaceted and includes planning for outbreaks; conducting multidisciplinary research across species; developing new vaccines and rapid diagnostic tools; effectively using the broad capabilities of university, industry, state veterinary diagnostic laboratories; and ensuring that an appropriate and state-of-the-art infrastructure exists to accomplish diagnosis. More than ever, there is a need to develop strong connections between public health and animal health officials, both domestically and internationally, and to expand the scope of animal disease concern to include wild and exotic animals.

4

Gaps in the Animal Health Framework

INTRODUCTION

The study's Statement of Task (Box 1-1) charges the committee with identifying key opportunities and barriers to the successful prevention and control of animal diseases. In its analysis of the existing animal health framework, as presented in Chapter 2, and the lessons of specific diseases and disease outbreaks, as presented in Chapter 3, the committee explored the responsibilities and actions of producers, regulators, policymakers, and animal health care providers and their effectiveness in providing disease prevention and control. Based on that review, the committee found that the main barriers to successful prevention of animal disease are gaps in the animal health framework that make it vulnerable to future animal disease threats, particularly from exotic animal diseases. The key gaps, identified in this chapter, are organized into the following categories:

- Coordination of Framework Components
- Technological Tools for Preventing, Detecting, and Diagnosing Animal Diseases
- Scientific Preparedness for Diagnosing Animal Diseases: Laboratory Capacity and Capability
- Animal Health Research
- International Issues
- Addressing Future Animal Disease Risks
- Education and Training
- Improving Awareness of the Economic, Social, and Human Health Effects of Animal Diseases

COORDINATION OF FRAMEWORK COMPONENTS

Gap 1: **A key gap in preventing, detecting, and diagnosing animal and zoonotic diseases is the lack of timely, appropriate, and necessary coordination and leadership among USDA, DOI, DHS, HHS, animal industries, and other responsible federal, state, and private entities.**

Whether due to historic structures and functions of the USDA, HHS, and related federal, state, and local governments, or because of changes and challenges in funding and resources, there is an apparent disconnect between agencies that should function in partnership. Examples of disease events, whether an emergent disease (monkeypox, West Nile virus, severe acute respiratory syndrome), endemic disease (chronic wasting disease and avian influenza), or exotic disease (foot-and-mouth, exotic Newcastle disease) reveal a lack of effective cooperation among local, state, and federal entities.

The Trust for America's Health (Benjamin et al., 2003) found over 200 different government offices and programs engaged in the response to just five outbreaks of animal-borne diseases (monkeypox, West Nile virus, bovine spongiform encephalopathy, Lyme disease, and chronic wasting disease). It also found that as many as seven cabinet-level agencies and hundreds of state and local public health agencies are involved. State departments of agriculture and environmental protection agencies also play critical roles. Table 4-1 shows the complexity (and the need for coordination and communication) of responsibility and the number of federal government agencies involved with each specific disease/agent examined in Chapter 3. The table does not show the significant overlaps (for diseases such as monkeypox) in the programmatic functions performed by various federal agencies that also exist. Whereas there are clear lines of responsibility and authority for exotic disease agents, such as highly pathogenic avian influenza, the regulatory lines of authority are not defined for endemic agents such as low pathogenic avian influenza, therefore hindering the nation's ability to prevent the potentially devastating spread of a disease before it develops. Furthermore, the system as a whole lacks integration, not only within the federal system, but also among federal agencies and programs directed through states or animal health organizations.

In addition, despite the number of federal agencies responsible for aspects of animal health policy (as shown in Table 4-1), there is a lack of federal oversight of the animal-centered aspects of zoonotic diseases. The monkeypox outbreak revealed no equivalent federal responsibility and only a limited federal animal health infrastructure for addressing a zoonotic disease outbreak transmitted by nonlivestock species.

Economic environments, social structures, and management practices are unique to different regions, requiring flexibility and tailored responses

TABLE 4-1 Primary Federal Jurisdictions for Specific Animal Diseases

Disease	Animals Affected	Government Agency(ies)
Exotic Newcastle Disease	Multiple avian species	DOI: NWHC USDA: APHIS, ARS, CSREES DHS: Bureau of Customs and Border Protection USTR OSTP
Foot-and-mouth Disease	Cattle, swine, and other cloven-hoofed species (sheep, goats, deer)	DOI: NWHC USDA: APHIS, ARS, CSREES DHS: Bureau of Customs and Border Protection DoD USTR White House: OSTP, OMB
Monkeypox	Prairie dogs, humans	HHS: FDA, CDC DHS: Bureau of Customs and Border Protection DOI: FWS USDA: APHIS, CSREES
Bovine Spongiform Encephalopathy	Cattle	HHS: FDA, CDC, NIH USDA: APHIS, FSIS, ARS, CSREES DHS: Bureau of Customs and Border Protection DOS DoD USTR White House: OSTP, OMB
Chronic Wasting Disease	Elk, mule deer, white-tailed deer	HHS: FDA DOI: BIA, NPS, FWS, BLM USGS: National Health Lab USDA: APHIS, ARS, CSREES EPA DoD
West Nile Virus	Mosquitoes, birds, humans	DOC DOI DoD EPA USDA: CSREES HHS: CDC, FDA, NIH
Avian Influenza	Avian species, humans, swine, cats	HHS: CDC USDA: APHIS, ARS, CSREES DHS: Bureau of Customs and Border Protection DOI: NWHC
Severe Acute Respiratory Syndrome	Humans, palm civets and raccoon dogs	HHS: CDC, FDA USDA: CSREES DOI

to ensure compliance and cooperation in prevention and early detection of disease events. In the current system, there is underutilization of private industry, local, and regional resources, as well as a reluctance to capitalize on expertise located outside of federal agencies. A wealth of scientific expertise in academic institutions and state diagnostic laboratories could be called on to provide expert advice and assistance. Timely communication with those on the front lines could be enhanced to improve disease detection and response. In the case of monkeypox, local stakeholders waited for instructions from the federal government, which hindered their ability to react.

TECHNOLOGICAL TOOLS FOR PREVENTING, DETECTING, AND DIAGNOSING ANIMAL DISEASES

Gap 2: **Efforts for the rapid development, validation, and adoption of new technological tools for the detection, diagnosis, or prevention of animal diseases and zoonoses are lacking or inadequate.**

New scientific tools and technologies with proven potential have not moved quickly into routine use within the current animal health infrastructure. State-of-the-art scientific approaches and technologies, often developed by and for basic research and military application, are often rapidly adopted by first-responder and public health communities to protect human health, but they have been significantly slower to transition into the animal health arena. Translational (or applied) research and federal funding sources to support the development, validation, and/or implementation of technological tools specifically for animal health applications are limited, and economic incentives for the private sector do not traditionally support these development efforts. In short, society will pay more to learn how to protect human health than to protect animal health. Federal and state laboratories across the country vary greatly in their ability to obtain advanced technologies, such as robotics for surge capacity, instrumental analyses (i.e., gas chromatography, mass spectrometry) for high-resolution toxin and protein detection, and molecular-based tools for rapid and sensitive agent detection or identification. Recent awareness and initiatives related to bioterror preparedness and homeland security have improved both federal and state laboratory access to rapid molecular-based diagnostic tools; however, as an industry, animal health lags behind the military, first-responder, and public health community in the use of field-based air sampling techniques, handheld devices, and similar technologies.

As demonstrated by the exotic Newcastle disease outbreak, the existing animal health framework in California was forced to develop and validate an effective detection tool during the outbreak, though awareness of the threat and the technology used had existed for many years. The delays in evaluating and implementing new and emerging technologies are also a concern with preventive vaccine strategies. Prevention strategies, including vaccines carrying markers that would allow a laboratory to distinguish a vaccinated animal from one exposed to a naturally occurring pathogen, have not been promoted in the United States. The lessons identified and reported from the 2001 U.K. foot-and-mouth disease (FMD) outbreak clearly indicate the importance of marker vaccines and/or diagnostic tests able to distinguish between an animal vaccinated against a foreign animal disease and an animal naturally exposed as a critical disease control strategy. Yet, little or no progress has been made in this direction in the United States, despite the availability of the technological tools. Unanswered questions remain about the adequacy of the supply and serotypes of FMDV vaccine available to the United States in the event of an outbreak (inadvertent or bioterrorist) and the surge capacity and biosafety level 3 (BSL-3) vaccine manufacturing facilities for rapid production of additional USDA-licensed doses.

Although basic research has generated prototype bioengineered animal vaccines, the translational research for their cost-effective production and testing in safety and efficacy field trials is lacking. Further, BSL-3 facilities to conduct translational vaccine research and to manufacture vaccines for BSL-3 pathogens are extremely limited in the United States.

Approaches to induce mucosal as well as systemic immunity and other preventive methods to block shedding of pathogens in diverse host species, including wildlife and companion animals, are largely undefined and illustrate the need for comparative medicine studies. The lack of an adequate understanding of safe and effective immune system modulators (adjuvants) to stimulate the various immune responses has resulted in few USDA approvals of adjuvants for use in animals. Stimulation of innate immunity is envisioned as a new strategy to achieve early initial nonspecific immune responses to pathogens; this approach could reduce pathogen load prior to specific vaccine response, shedding, and transmission in animal populations in the face of an exotic, zoonotic, or other pathogen exposure.

SCIENTIFIC PREPAREDNESS FOR DIAGNOSING ANIMAL DISEASES: LABORATORY CAPACITY AND CAPABILITY

Gap 3: **The animal disease diagnostic system in the United States is not sufficiently robust to provide adequate capacity and capability for early detection of newly emergent, accidental, or intentionally introduced diseases.**

More specifically, the committee found significant delays in development and validation of diagnostic assays and proficiency testing programs; inadequate use and coordination of the diagnostic resources and expertise located throughout the country; insufficient integration and coordination of laboratory diagnosis of zoonotic and foodborne diseases; limited surge capacity; inadequate diagnostic BSL-3 biocontainment; and limited research on and implementation of new diagnostic tools and methodologies.

Laboratory diagnosis of animal diseases in the United States is multifaceted and involves federal, state, and commercial entities. However, the committee's review focused primarily on assessing publicly funded laboratories and the current operational status of national laboratory networks. While commercial laboratories are an important part of the system, we did not conduct an in-depth analysis of this component.

Funding and implementation of the National Animal Health Laboratory Network (NAHLN) was a major step and is an important and necessary paradigm shift from an exclusive federal to a shared responsibility for foreign animal disease diagnosis (see Chapter 2). However, the current network does not provide the necessary surge capacity and is not prepared for disease agents and toxins outside the narrow list of eight exotic diseases[1] that provided an initial focus for network development. Furthermore, coordination among federal and state laboratory networks is not optimum. The committee reviewed the current status of veterinary diagnostic laboratory membership in the Laboratory Response Network for Bioterrorism (LRN) and concluded that there is limited and insufficient linkage among veterinary diagnostic laboratories, the NAHLN, and the LRN to respond to agents with zoonotic concerns, such as monkeypox or West Nile virus (see Chapter 3).

The committee reviewed the current status of BSL-3 diagnostic laboratory and necropsy space in the United States and found a significant deficiency in this capital resource. Since *all* exotic animal diseases are classified as BSL-3 agents, and since most of the OIE category A select agents are zoonotic and require the handling of known or potentially infected animals and animal-derived tissues in veterinary diagnostic laboratories under BSL-3 containment, the current capacity and distribution of BSL-3 space for necropsy (autopsy) of animals and laboratory workups are insufficient for routine diagnosis and grossly insufficient for surge capacity. It is possible that accidental or intentional introduction of these agents could occur in any state or region, and it is unlikely that movement of

[1]The eight selected agents are avian influenza, exotic Newcastle disease, foot-and-mouth disease, classical swine fever, rinderpest, bovine pleuropneumonia, African swine fever, and lumpy skin disease.

carcasses or shipment of specimens will be time-sensitive enough to rely on neighboring regions for BSL-3 laboratory capacity.

In addition to reviewing the diagnostic networks as a component of disease prevention, detection, and diagnosis, the committee analyzed the approach toward laboratory diagnostics currently practiced in the United States. The committee found that the traditional approach, which focuses mainly on individual animal diagnosis and generally does not formally consider population diagnosis, is not addressing the need for population-based diagnostic methods and information that are critically important in identifying the multicausality of disease and the factors predisposing or contributing to development of new or emerging diseases.

ANIMAL HEALTH RESEARCH

Gap 4: **The nation supports only limited multidisciplinary research to address prevention and detection of animal disease (both zoonotic and nonzoonotic) by studying factors related to pathogenesis, interspecies transmission, epidemiology, and ecology.**

Early recognition of emerging diseases requires a fundamental knowledge of the epidemiology of the disease, which includes an understanding of the agents and hosts in their natural environment. The interactions of the intrinsic and extrinsic factors that lead to emergence of new disease are poorly understood. Many individual researchers address various diseases relatively independently and usually with a focus on a single host species or mouse model. As a result, medical scientists may be unaware of key research done in other species by veterinary scientists studying a similar, closely related, or even unique animal pathogen. For example, the animal reservoir and susceptible species for SARS remain undefined and integrated, and collaborative research efforts to study the responses to SARS in infected humans and diverse animal hosts have not been instituted in the United States. Furthermore, basic and translational research related to prevention, detection, and diagnosis of animal and zoonotic diseases is conducted through an array of government, academic, and private institutions; however, no mechanism exists to coordinate research dollars and priorities to ensure that important topics are not overlooked and to ensure the most effective use of existing research dollars.

As demonstrated by SARS and many other disease outbreaks, research ties, interagency funding, and cooperation for shared research between biomedical and veterinary scientists are lacking. Opportunities to train biomedical and veterinary public health personnel in infectious and zoonotic disease research are limited. Funds are lacking to study disease pathogenesis in appropriate animal hosts and to investigate zoonotic diseases including identifying animal reservoirs and the mechanisms and chains of interspecies transmission. The current emphasis of major federal funding agencies supporting research related to animal health or zoonoses (e.g., USDA, NIH, NSF) focuses on basic research, providing less opportunity for translational research aimed at the immediate goals of prevention (e.g., vaccines, antimicrobials, producer behavior), detection, and diagnosis of animal and zoonotic diseases. Development and validation of veterinary diagnostic assays utilizing state-of-the-art molecular techniques has not been a priority for federal funding prior to homeland security interests in protection against agricultural bioterrorism. Existing efforts remain limited and focused on assay development for biothreat agents and foreign animal diseases. While the threat of bioterrorism and the recent epidemics of zoonoses, such as SARS, West Nile virus, and monkeypox, have boosted the priority for this type of translational research, there is little coordination among agencies and a continuing deficit of funding to scientists outside the federal arena. In addition, a robust national system for considering validation data on new assays and for adopting validated assays in one species to other species and matrices does not exist. There is currently limited emphasis on multidisciplinary research on disease pathogenesis, interspecies transmission, or comparative medicine.

Gap 5: **There are not enough biosafety level 3 (BSL-3 or BSL-3 Ag) facilities and those that exist are not strategically located throughout the United States. Not all level 3 facilities are suitable or equipped for research on diseases (including zoonoses) of livestock, poultry, or wildlife requiring that level of biocontainment.**

When a new infectious agent is suspected, initial efforts must be directed to the rapid definition and characterization of the agent, as was done during the SARS outbreak (described in Chapter 3). One of the first actions is to contain the agent and to characterize it under strict biocontainment conditions. Subsequent research requires assessment of the pathogenic potential, origin and host range, pathogenesis, diagnosis, and preventive or control measures in the natural and susceptible host species, including large domestic animals and wildlife. State-of-the-art equipment and technological tools to conduct the research needed to understand, prevent, and control these emerging or exotic infectious agents are not widely available.

Both the monkeypox and SARS outbreaks helped to identify weaknesses in the veterinary laboratory infrastructure in the United States, which does not have appropriate biocontainment facilities to allow investigative research on potential animal hosts and disease transmission. Preliminary data from a recent survey conducted by the American Association of Veterinary Medical Colleges indicate that federal and state BSL-3 large animal biocontainment facilities in the United States are limited to two states, with an additional five ABSL-3 facilities under construction that will be capable of holding large animals. Only one ABSL-3 Ag facility currently exists in the United States (Plum Island), and only two of the facilities currently under construction will be ABSL-3 Ag biocontainment (Richard Dierks, AAVMC, personal communication, November 2004).[2]

INTERNATIONAL ISSUES

International Interdependence and Collaboration

***Gap 6:* The United States is not sufficiently engaged with international partners to develop strategic approaches to preventing, detecting, and diagnosing animal diseases before they enter this country.**

International collaborations are largely ad hoc, resulting in a nonstrategic system for dealing with global animal health issues.

Global trade, population, and production in other countries, along with advancements in technical capacity and our own regulatory infrastructure, are some of the factors eroding the historically strong trade status enjoyed by the United States. Adding to these global factors are international standards and legally binding trade agreements such as the World Trade Organization (WTO) Agreement on Sanitary and Phytosanitary Standards (SPS), formulated with technical expertise from different countries. Every effort is made to adopt standards by consensus, and each country, no matter its size or other qualifying factors, is on equal footing with one vote. Adoption of standards requires the efforts and encouragement of all countries to comply and provide timely notification of zoonotic and exotic diseases. To operate independently is no longer a viable option for the United States, despite its large stake in the global economy and the world. Moreover, although the United States has great economic and political influence, it constitutes a small fraction of the total world population. (World population growth every 3–4 years equals the entire population of the United States.)

---

[2]The committee sought information on the extent of BSL-3 biocontainment facilities for large animals in the United States. Definitive information could not be supplied. However, at the time the report was finalized, the AAVMC had obtained preliminary results from a survey of BSL-3 facilities in veterinary medical colleges and departments in the United States. The information provided is based on preliminary results from AAVMC's November 2004 survey.

Importation, Sale, and Transport of Exotic Animals

Gap 7: **The current patchwork of federal policies and agencies with limited or ill-defined jurisdiction for the import, sale, and movement of exotic and wild-caught companion animals and zoo specimens is a significant gap in preventing and rapidly detecting emergent diseases.**

As the monkeypox outbreak revealed, there is no defined federal responsibility beyond that for protecting public health, and only an informal federal animal health infrastructure, for addressing a zoonotic disease outbreak transmitted by a nonlivestock, nonwildlife species. Prior to the interim final rule (banning the import, sale, or distribution of prairie dogs and some African rodents responsible for the monkeypox outbreak), import and movement of exotic animals was uncontrolled. Some states have bans on sale and distribution of prairie dogs to prevent transmission of plague. However, no uniform federal regulations that would provide equivalent controls nationwide have been established. Tracking of these animals in the United States is inconsistent and ineffective, and there is a disturbing lack of standardized testing of the health status of exotic animals at the point of origin or importation, in companion animal shops, at trade fairs, and in other venues. Regulatory authority for the intrastate movement of these animals once they are in the United States lies with the states. However, state infrastructure to oversee and effectively monitor movement of nonlivestock species is inconsistent and often weak, with neither the budget nor the personnel within most states. Exotic animals are imported daily into the United States with little or no health monitoring, increasing the probability of an animal pathogen or zoonotic disease entering the United States through an imported animal such as occurred with human salmonellosis via turtles or monkeypox through African rodents. Wild animals transported from their native habitat and introduced into live animal markets may harbor unknown disease agents transmissible to humans such as monkeypox, influenza, and perhaps SARS, which is highly likely to also be a zoonotic disease (as described in Chapter 3).

ADDRESSING FUTURE ANIMAL DISEASE RISKS

Gap 8: **The current animal health infrastructure for food-animals, wildlife, and companion animals does not have formal and comprehensive-based science and risk analysis systems for anticipating potential challenges to animal health; ranking their likelihood of occurring and likely severity; evaluating alternative prevention, detection, and diagnostic systems; and using this information to make appropriate policy decisions.**

The traditional approach for disease prevention in the United States has been to formulate disease control strategies based on whether the specific disease is present or absent, rather than relying on a risk-based approach for developing disease control strategies. For infectious and economically devastating animal diseases such as FMD, the United States has made substantial investments in disease eradication and historically has moved to restrict trade from countries where the disease is reported if it poses risk to the health status of the U.S. population or jeopardizes U.S. export markets. In the past, the United States adopted a zero-risk policy for disease introduction as the most expedient approach, though this relied on a strong infrastructure with financial resources to rapidly respond should the disease be inadvertently or intentionally introduced. On the other hand, trade partners intrinsically link the economic viability of entire export markets with the absence or presence of disease within their borders, creating issues of compliance and international cooperation. Trade restrictions, and even perceptions of unfair regulation, have created significant uneasiness among trading nations, as witnessed by markets and political responses in Europe, Asia, Canada, and the United States following respective initial detection of BSE.

EDUCATION AND TRAINING

Gap 9: **There is an inadequate supply of veterinarians educated for careers in research, public health, food systems, ecosystem health, diagnostic laboratory investigation, and rural and/or food animal practice.**

The face of veterinary medicine has changed over the years from a focus on rural practice to a profession dominated by practitioners serving small companion animals. There are not sufficient graduates to meet the needs in a number of major and distinct fields of veterinary medicine dealing with various species of food animals, rural practice (mixed domestic animals), ecosystem health (including wildlife and conservation medicine), public health, the many dimensions of the food system, and biomedical science. In addition, veterinary graduates are not adequately prepared to deal with foreign animal diseases, public health (Hoblet et al., 2003; Walsh et al., 2003), the food system (Hird et al., 2002), ecosystem health (Van Leeuwen et al., 1998), and biomedical research, without further postgraduate education. According to the Association of American Veterinary Medical Colleges (AAVMC), the 28 veterinary colleges in the United States graduate approximately 2,300 veterinarians per year, and currently they cannot keep up with societal needs in private or public practice (AAVMC, 2004).

The committee also found a steady decline in the number of rural practitioners and of veterinarians employed in regulatory agencies, due to consolidation in the animal production sector and decreases in extension education and research. Today, veterinary capacity in rural areas is largely composed of consulting veterinary businesses and limited numbers of mixed animal practices. These decreases threaten to undermine the nation's capacity to protect animal and human health and to respond to potential national emergencies. The USDA, presently underserved, predicts a shortfall of 584 veterinarians by 2007. Fifty percent of U.S. Public Health Service veterinarians are currently eligible for retirement (AAVMC, 2004).

Furthermore, the changing emphasis toward and greater specialization in companion animal private practice has created a critical gap in the current animal health infrastructure. Critical areas—including food safety, emerging and foreign animal diseases, public health, food systems, and animal agriculture in general—are no longer being adequately addressed. This gap is further accentuated by the fact that training and continuing education are now primarily focused on companion animal practice which, in turn, has reduced the overall awareness and importance of these critical needs and weakened the current animal health infrastructure. Thus inadequate veterinary capacity is a growing problem from the perspective of its distribution, total number, and range of competencies. It is beyond the scope of this report to undertake an in-depth analysis of veterinary shortages, as described in the NRC report *National Need and Priorities for Veterinarians in Biomedical Research* (NRC, 2004b). However, the impact of this shift in the profession away from rural animal practices and public service sectors has a profound impact on the recognition and early detection of foreign animal diseases.

***Gap 10:* Education and training of those on the front lines for recognizing the signs of animal diseases is inadequate.**

Animal handlers and people working and living with animals on a day-to-day basis form the true first line of detection for animal diseases. Producers and animal personnel may not report suspicious signs to the herd veterinarian or government official simply because they do not know what the signs look like, or do not recognize them and understand the implications of delays in recognizing and reporting signs of disease. Each year, only about 250 to 300 foreign animal disease investigations are submitted to the Foreign Animal Disease Diagnostic Laboratory (FADDL) (Tom McKenna, Plum Island Animal Disease Center, personal communication, June 2005). As noted in Chapter 3, veterinary oversight of animal units in the United States is hit-and-miss, with progressive, health-minded producers more likely to engage veterinary services to provide education and training and to promote observation for animal diseases and early reporting. Veterinarians also may not be able to readily recognize a foreign or exotic disease when examining animals with clinical diseases that mimic common indigenous diseases. Furthermore, animal handlers and other personnel working directly with animals may be inadequately educated and trained in the detection of foreign and exotic animal diseases.

The committee acknowledges the critical need for professionals in human medicine, wildlife health, and public health, as well as domestic animal health, to identify, recognize, and report cases of zoonotic diseases within their respective areas of expertise and responsibilities. Although it was beyond the capacity of its review, the committee was nevertheless very concerned about the level of knowledge and understanding of zoonoses and especially emerging zoonoses within the various health protection communities.

IMPROVING PUBLIC AWARENESS OF THE ECONOMIC, SOCIAL, AND HUMAN HEALTH EFFECTS OF ANIMAL DISEASES

Gap 11: **Despite the vital role that animal health plays in preserving the safety of the food supply and other important societal resources, there is little national consumer awareness or public investment in maintaining a viable animal health infrastructure to protect and defend this critical resource.**

As described in Chapter 3, the recent outbreaks of foot-and-mouth disease in the United Kingdom; of SARS in Asia and Canada; and of exotic Newcastle disease, avian influenza, and BSE in the United States are reminders of the threats such diseases pose to the U.S. food supply (as well as confidence in the safety of the food supply), the global economy, and public health (Shadduck et al., 1996). Today the entire food and fiber system—including farm inputs, processing, manufacturing, exporting, and related services—is one of the largest sectors of the U.S. economy and accounts for output of $1.5 trillion, or nearly 16 percent of the gross domestic product, and 17 percent of the civilian labor force (USDA, 2003).

The annual value of livestock production (cash receipts) was nearly $100 billion during the 1993-2002 period, about half of the total value of agricultural sector production (McElroy et al., 2003). The United States exported nearly $55 billion in agricultural exports in 2002, with animals and animal products accounting for over 20 percent (USDA-ERS, 2001). Wildlife and companion animals also have significant recreational and environmental value. Economic activity based on wildlife-related recreation in 2001 in the United States was estimated to be $108 billion. In 2005, expenditures on pets are projected at a record-high of $35.9 billion (APPMA, 2005).

Given the economic value of animals, a key question is whether sufficient resources have been allocated to safeguard animal health; in other words, to prevent disease outbreaks from occurring. Although not all disease events are catastrophic on a large scale, one analysis suggests that

3–4 percent of the value of animal production in agriculture is routinely lost to animal diseases (Hennessy et al., 2005) including the cost of prevention. Another estimate suggests that the losses might be even higher: up to 18 percent of the annual farm gate value of animal commodities, costing production agriculture and the U.S. economy billions of dollars each year (FAIR, 2002). A study of livestock diseases in the United Kingdom estimates a range in the costs of losses from disease relative to the costs of treatment and prevention measures (Bennett, 2003). In that study, mastitis in dairy cattle ranked the highest in terms of direct costs from losses (over £120 million or nearly $200 million dollars, in 1996 values), while prevention expenditures were estimated to be £4 million, or $6.6 million dollars .

The financial investment to prevent disease may be far less than the losses in value should a disease occur, particularly when losses include not only the cost to producers and associated industries, but also the value of social welfare such as the loss of food and other products, or the loss of a companion or zoo animal.

However, for several reasons, the general public is unaware of the full costs of disease. First, a large and growing percentage of the general population is increasingly removed from a basic understanding of agriculture, its links with animal health, and related sectors (Whitener and McGranahan, 2003). A lack of personal experience or knowledge about animal production may lead the general public (consumers) to undervalue efforts required to prevent animal diseases, or to recognize that losses from disease may be reflected in higher costs for food, recreation, or health care. When the public is not aware of these costs, they (consumers, business) will underestimate the value of prevention, detection, and diagnosis (Colorado State University and Farm Foundation, 2003; Sumner, 2003).

The public understanding of the purposes and the need for animal research and for disease prevention measures might also be affected by societal attitudes toward animals that have been fostered by animal activists. Public education and ongoing risk communication with the general public improve the ability of consumers to make appropriate decisions and build support for national animal disease management efforts. Research on effective methods and tools of risk communication would make an important contribution to building an effective animal health infrastructure.

SUMMARY

The current animal health framework was built on animal management practices, economic impacts, and societal norms that are no longer valid. At the same time, animal and human populations and their interfaces have changed and continue to change. The committee analyzed the current capabilities and limitations of the animal disease framework and identified the following 11 gaps that hinder effective prevention, detection, and diagnosis:

- There is a lack of timely, appropriate, and necessary coordination and leadership among USDA, DOI, DHS, HHS, animal industries, and other responsible federal, state, and private entities.
- Efforts for the rapid development, validation, and adoption of new technological tools for the detection, diagnosis, or prevention of animal diseases and zoonoses are lacking or inadequate.
- The U.S. animal disease diagnostic system is not able to provide adequate capacity and capability for early detection of newly emergent, accidental, or intentionally introduced diseases.
- The nation supports only limited multidisciplinary research to address prevention and detection of animal disease (both zoonotic and non-zoonotic) by studying factors related to pathogenesis, interspecies transmission, and ecology.
- The number of BSL-3 facilities is inadequate, and the existing labs are not strategically located in the United States nor are they suitably equipped for research on diseases requiring biocontainment.
- The United States is not sufficiently engaged with international partners to develop strategic approaches to preventing, detecting, and diagnosing animal diseases before they enter this country.
- Federal policies and agencies have limited or ill-defined jurisdiction for the import, sale, and movement of exotic and wild-caught companion animals and of zoo specimens, creating a loophole, allowing a significant gap in preventing and detecting emergent diseases.
- The nation lacks a formal and comprehensive-based science and risk analysis system for anticipating potential challenges to animal health and for use in policy decisions.
- The supply of veterinarians in research, public health, food systems, ecosystem health, diagnostic laboratory investigation, and rural and/or food-animal practice is inadequate.
- Education and training of those on the front lines for recognizing the signs of animal diseases is inadequate.
- Little national consumer awareness or public investment in maintaining a viable animal health infrastructure exists.

Based on these gaps, the next chapter provides key opportunities to strengthen the framework to successfully prevent, detect, and diagnose animal diseases.

5

Recommendations for Strenghtening the Animal Health Framework

INTRODUCTION

As the committee reviewed a group of contemporary zoonotic disease issues, it became apparent that the convergence of human and animal diseases has also produced a convergence of public health and animal health officials, experts, and organizations locally, nationally, and globally. The critical lesson identified from West Nile virus, severe acute respiratory syndrome, monkeypox, bovine spongiform encephalopathy, and other recent zoonotic events is the need for a new strategic partnership and relationship between the public and animal health communities. This critical need represents a significant gap that should be filled. In this era of emerging and reemerging zoonoses, there is a distinct lack of coordination and development of strategic partnerships between the human and animal health communities. It is also a reminder that veterinary medicine is one of the health professions and, as such, is committed by oath to the improvement of public health.

It has been asserted that 11 of the last 12 significant human epidemics have been due to zoonotic pathogens (Torrey and Yolken, 2005). Considering the factors that have led to this new era, there is nothing to suggest that this trend of emerging and reemerging zoonoses will abate. Consequently, the animal health framework has new responsibilities and obligations to address the serious and profound occurrence of this group of diseases.

The challenges and opportunities now faced by the animal health framework are described throughout this report and are reflected in the committee recommendations presented here. Future research, training, laboratory linkages, global planning, diagnostics, and other features pertaining to the prevention, diagnosis, and detection of zoonoses now deserve urgent attention and should be part of the cornerstone and culture of the evolving animal health framework. Strategic partnerships in human health and animal health will likely be fundamental to the future success of both communities.

COORDINATION OF FRAMEWORK COMPONENTS

Recommendation 1: **The nation should establish a high-level, centralized, authoritative, and accountable coordinating mechanism or focal point for engaging and enhancing partnerships among local, state, and federal agencies and the private sector.**

A centralized coordinating mechanism or focal point could help ensure flexibility and strong coordination at the federal, state, and local levels; help minimize duplication of effort; and maximize efficiency in both resource allocation and function.

A few examples of possible means to accomplish this mechanism could be through a high-level individual or through a group embedded in an existing office or an interagency alliance. Alternatively, given the overlapping legislative mandates and competition for resource allocations that affect federal agencies and issues of state versus federal jurisdiction and authority, a third approach would be to establish a nongovernmental organization to serve as a central coordinating agency. Such an organization would function like a domestic version of the World Organization for Animal Health (OIE), which arbitrates between governments. Another example of an organization that has some of the properties and function that a central coordinating body should have is the Southeast Cooperative Wildlife Disease Study (SCWDS), a state-federal cooperative agency whose resources are supported and shared by the wildlife agencies of 15 states and Puerto Rico, the U.S. Department of the Interior, and U.S. Department of Agriculture (USDA). The SCWDS is a multiple-purpose program that carries out research, performs diagnostics and surveillance of wildlife diseases, and provides training and consultation to wildlife managers, farmers, landowners, veterinarians, physicians, and the agencies tasked with safeguarding animal health.

In making these suggestions, however, the committee wishes to clarify that it was not tasked to analyze all possible mechanisms for implementing Recommendation 1. Furthermore, while there is compelling evidence of a need for improved coordination in prevention, detection, and diagnosis, the committee felt it is premature to recommend a specific systemwide mechanism prior to examining other parts of the animal health framework—that is, surveillance, monitoring, and response recovery. The two planned successor phases of this study can potentially better address this issue following examination of the other aspects of the animal health framework.

Regardless of how a central coordinating mechanism or focal point is implemented, it will need to promote effective communication among various stakeholders and with the public during and outside times of animal disease outbreaks. Opportunities for information sharing between agencies using electronic information systems should be developed. A methodical effort should be made to identify and link key databases and establish protocols for contributing data and for creating alerts. For example, subject matter experts in government, industry, and academia will sometimes need access to information gathered by the intelligence community. In addition, public-private partnerships are important in long-term strategic planning. The private sector will always be the beneficiary of effective animal disease prevention, detection, and diagnostic programs, but much of the success will depend on the level of private sector leadership, involvement, and investment.

While the framework will promote effective communication and collaboration among different stakeholders, it will also need to ensure that those on the front lines of disease prevention and detection at the local level (e.g., field personnel and wildlife management) are fully integrated into the system of animal health communication and of disease prevention and detection. As was demonstrated during the exotic Newcastle disease outbreak, private industry, local and regional resources, and a willingness to capitalize on expertise located outside the centralized federal animal health system allowed a cost-effective and reliable assay to be rapidly developed, validated, and implemented for disease detection and control. Another specific example of broad-based local community involvement in animal disease prevention and detection is described in Box 5-1.

TECHNOLOGICAL TOOLS FOR PREVENTING, DETECTING, AND DIAGNOSING ANIMAL DISEASES

Recommendation 2: **Agencies and institutions— including the U.S. Department of Agriculture (USDA) and the Department of Homeland Security (DHS)— responsible for protecting animal industries, wildlife, and associated economies should encourage and support rapid development, validation, and adoption of new technologies and scientific tools for the prevention, detection, and diagnosis of animal diseases and zoonoses.**

The current animal health framework has been slow to evaluate, validate, and implement new scientific tools and technologies that could sig-

BOX 5-1
Example of Preparedness, Prevention, and Detection Plan of Action

Wichita County, Kansas, provides an example of a broad-based, local community prepared to prevent or detect an animal disease outbreak (intentional or nonintentional). Social and public health professional representatives of law enforcement, fire, emergency medical services (EMS), county commission, chamber of commerce, county extension, and Farm Service Agency joined the animal health community of veterinarians, pharmaceutical representatives, livestock producers, and others interested and potentially impacted in the economy. Coordinating this widespread committee was first prompted by the threat of agroterrorism or a foreign animal disease outbreak. Many representatives had never met each other and were not aware of one another's concerns and issues. Over 6 months, the committee developed and implemented a preparedness emergency plan for the county that subsequently became the model for all counties in the state.

Major components of the plan are:

- List of vulnerable entities in each county such as feedlots and livestock markets with names and numbers to contact in case of an emergency.
- List of resources such as feedlots with loaders, cooperatives with fencing materials and feed, and volunteer agencies that would take care of feeding and housing of workers.
- All entities and agencies involved.
- All emergency team members' names and contact information.

nificantly enhance animal disease prevention, detection, and diagnostic capabilities for the United States. Despite a recent surge in activity related to post-September 11 homeland security efforts and associated focused funding, the active review and implementation of advancing technologies is inadequate to protect and enhance the health of the country's animal populations and related economies. Existing technological advances, such as immune system modulators, animal-embedded monitoring (chips embedded underneath an animal's skin to monitor temperature and other physiological indices), and differential vaccines as prevention strategies, as well as a range of rapid, automated, sensitive, and portable sampling and assay systems for early warning and reliable diagnosis, are not adequately exploited by the current animal health framework. Early biodefense warning systems, such as DHS's BioWatch or private industry's

- Notification procedures - courses of action to be taken and implemented on short notice.
- Procedure for emergencies.

Strengths and Challenges of the Coordinated Local Emergency Preparedness Plan:

- Each community sector has knowledge of the diverse capacities and capabilities of the overall system.
- Each committee member is a liaison for his/her respective agency or entity and is empowered to speak for his/her organization.
- All committee representatives are holistically and responsibly involved with the coordination of the plan.
- Periodic review of the County Preparedness Plan and participating agency/entity meetings with "scenario" exercises are critical to keeping the plan fresh and current and in keeping the Plan updated.
- Dispersing prudent reliable information to the media and the public is extremely important.
- Education of county residents (particularly agricultural) is important, and should be an ongoing activity, with regard to the care and handling of livestock. Trained and educated livestock producers will utilize best management practices (BMPs) in the prevention, detection, and diagnoses of animal disease.

Successful preparedness in the event of a foreign animal disease outbreak, intentional or nonintentional, can help eradicate disease and minimize adverse market reaction and economic impact.

gene-based anthrax testing, are designed for rapid detection and identification of key pathogens by sampling air in public environments. These systems have been operating since early 2003 and are meant to assist public health experts in rapidly responding to the intentional release of a biological agent (DHS, 2004a). Early warning technologies have not yet been adequately evaluated by the animal health infrastructure.

SCIENTIFIC PREPAREDNESS FOR DIAGNOSING ANIMAL DISEASES: LABORATORY CAPACITY AND CAPABILITY

Recommendation 3: **The animal health laboratory network should be expanded and strengthened to ensure sufficient capability and capacity for both routine and emergency diagnostic needs and to ensure a robust link-**

age of all components (federal, state, university, and commercial laboratories) involved in the diagnosis of animal and zoonotic diseases.

A robust animal health laboratory network will utilize the resources and expertise located throughout the country for efficiency in routine, early detection of agents, and provision of surge capacity in the event of disease outbreaks, whether newly arising, accidental, or intentionally introduced. Federal laboratories should serve as reference laboratories of the highest caliber and should rely on linkages through the network for assistance in assay development, standardization, and validation. The committee has identified five high-priority elements critical to the success of implementing this recommendation.

Based on the committee's review of the current operational status of the National Animal Health Laboratory Network (NAHLN) and the need for a more robust, broad laboratory network for optimum detection and diagnosis, *the first critical element is for the USDA, in collaboration with the American Association of Veterinary Laboratory Diagnosticians (AAVLD), to ensure that all accredited state/university diagnostic laboratories are members of the network, and that each of these laboratories is appropriately equipped and has trained staff to perform standardized assays developed and approved for diagnosis of high-consequence livestock and poultry pathogens and toxins.* This would ensure that all states with accredited laboratories could actively participate in the NAHLN at the reference level for multiple diseases and multiple assay formats, and not just by virtue of performing contract surveillance work for USDA. It would also ensure that diagnostic assays and systems meet rigorous scientific standards. While the committee agrees with the concept of including contract laboratories in the NAHLN, the original concept of building the laboratory system in the United States that ensures appropriate staffing, equipment, training, and use of standardized assays should not be lost or abandoned in this redefinition.

A second critical element is adequate quality assurance for network laboratories. There needs to be an independent process to plan, undertake, and evaluate proficiency testing for all laboratories involved in diagnosis and detection of animal diseases, including USDA, HHS, state, and other national laboratories. All network laboratories must meet standards for accreditation, verified by outside, independent review. In addition, USDA, in collaboration with its diagnostic partners, needs to develop and implement scientifically rigorous validation procedures for assessment of accuracy, precision, and utility of diagnostic assays and disease detection systems. An independent scientific review process should be utilized to evaluate and make recommendations for use and application of diagnostic and detection technologies and systems.

A third critical element is formalization of linkages and operational relationships among veterinary diagnostic laboratories, the NAHLN, and the Laboratory

Response Network for Bioterrorism (LRN). All fully accredited veterinary diagnostic laboratories that meet LRN membership requirements should be in the LRN. Membership of LRN reference laboratories that meet NAHLN membership requirements in the NAHLN also should be considered, though the committee did not examine this issue in sufficient depth to make a recommendation. The processes for laboratory membership in the NAHLN and the LRN are nearly identical, making joint membership nearly seamless. Joint membership could include sharing of assay protocols, proficiency testing programs, equipment, and staff as appropriate, and the networks could work toward standardized messaging for facile data exchange and mutually understood and approved reporting guidelines and requirements. The expected outcomes of this effort would include: an enhanced laboratory capability and capacity for detection and diagnosis of multiple agents (including zoonotic pathogens); standardized operating procedures for high-priority diseases that are shared and utilized by all laboratories in these networks; a common understanding and adherence to business practices and reporting procedures; and fluid electronic sharing of data in emergency and routine diagnostic scenarios. The committee recognizes that additional laboratory networks are under development, such as the Food Emergency Response Network, and encourages sharing of data with these laboratories as necessary and appropriate for prevention, detection, and diagnosis of animal diseases.

The 2002–2003 exotic Newcastle disease outbreak in California and neighboring states clearly showed the advantage of having high containment facilities at animal laboratories for tissue sampling and laboratory workup close to the outbreak location. *A fourth critical element of a strengthened laboratory infrastructure in the United States is expansion of BSL-3 laboratory and necropsy space in animal disease diagnostic laboratories for detection and diagnosis of agents requiring enhanced biological safety.*

A major paradigm shift from a focus on the pathology of individual animals or samples to a holistic diagnostic approach involving the population (herd or flock) is needed. Herd-flock diagnostics necessitate considering the multicausality of disease/health in animal populations, including the interactions among management practices, the environment, and infectious or toxic diseases. *A fifth critical element of a robust diagnostic system in the United States is the development and application of population-based diagnostic and detection methodologies.*

It will be necessary to improve education and, through research, develop cost-effective, efficient laboratory and field-based diagnostic strategies to diagnose and detect diseases, infectious agents, and toxins affecting herd or flock health. Diagnosticians and clinical epidemiologists are in a unique position for communication and instruction and should be encouraged to enhance producer and practitioner awareness and understanding of population-based approaches to diagnose and prevent disease, whether endemic or exotic. A broad diagnostic outlook for the herd or flock considering population health, rather than disease in only one or a few animals, and expanding diagnostic perspectives will increase the likelihood of early recognition of new or emerging diseases. These efforts will improve the overall chance of detecting a foreign animal disease through a broader accession base and increased interest at the farm level, in the wild, and in companion animals. Inherent in this new strategy of enhanced prevention and early detection is a means of funding routine diagnostic testing for indigenous diseases that mimic foreign or exotic animal diseases.

ANIMAL HEALTH RESEARCH

Recommendation 4: **Federal agencies involved in biomedical research (both human and veterinary) should establish a method to jointly fund new, competitive, comprehensive, and integrated animal health research programs; ensure that veterinary and medical scientists can work as collaborators; and enhance research, both domestically and internationally, on the detection, diagnosis, and prevention of animal and zoonotic disease encompassing both animal and human hosts.**

This process might be modeled on the National Institutes of Health (NIH)-administered Interagency Comparative Medicine Research Program, an interagency task force model, or some comparable process that promotes this type of cooperative research agenda.

This recommendation builds on a recommendation in the IOM report *Microbial Threats to Health: Emergence, Detection, and Response,* which states: "NIH should develop a comprehensive research agenda for infectious disease prevention and control in collaboration with other federal research institutions and laboratories (e.g., Centers for Disease Control and Prevention, Department of Defense, the U.S. Department of Energy, the National Science Foundation), academia, and industry" (IOM, 2003).

The agenda should include collaborative research among veterinary and medical scientists to provide an integrated research approach to detect and prevent zoonotic diseases infecting both human and animal hosts. In addition, it should include integrated research on comparative medicine to address interspecies transmission, disease pathogenesis, and host responses in diverse species including wildlife. Zoonotic strains of avian influenza, concerted research efforts, and dialogue involving veterinary and medical scientists are needed to address the most applicable control measures to be implemented in avian species to block transmission to humans. Research would not be limited to domestic activities only but would include the international dimensions, such as developing preven-

BOX 5-2
Government/Industry/University Research Partnership in Development of a Preclinical Test for Sheep Scrapie

In 1998, scientists at USDA's Agricultural Research Service, working in collaboration with researchers at the Washington State University College of Veterinary Medicine, the American sheep industry, and a private company, Veterinary Medical Research and Development, Inc., developed a preclinical, noninvasive test for scrapie. Scrapie is a degenerative and eventually fatal disease that targets the central nervous systems of sheep and goats. These researchers found that lymphoid tissue in the third eyelid of sheep collects prions, the unique protein that causes scrapie, bovine spongiform encephalopathy, chronic wasting disease, and other related diseases. They also designed a new antibody to identify prions in a sample of eyelid tissue (O'Rourke et al., 1998, 2000).

There is no cure or treatment for scrapie and scientists do not fully understand how it is transmitted. Sheep can harbor the disease for up to 5 years before they show signs such as trembling, incoordination, or scraping against objects. Prior to development of the eyelid test, diagnosis required recognition of clinical signs and testing of a sample of brain tissue collected from a euthanized animal. Producers with confirmed cases of scrapie in their flock had to destroy clinically normal animals to obtain the appropriate sample for a diagnosis in an effort to eliminate the disease. The ability to diagnose the disease at a much earlier stage has greatly facilitated attempts to eradicate it.

This partnership of government scientists, academia, a producer group, and private industry resulted in the first, and to date, the only validated preclinical test for a prion-induced disease. The test is approved by USDA APHIS and is being used in formal eradication programs in North America.

tion, detection, and diagnosis methods closer to the point of infection, and approaches that invite collaboration with other countries and related organizations.

One example of an effective research partnership between government, academia, and industry is described in Box 5-2.

The committee considered two possible mechanisms or strategies for implementing this recommendation: the establishment of an independent advisory group and/or an incentive program. The advantages and disadvantages of these alternatives are discussed below.

An independent scientific advisory group to the USDA (including ARS) with members from academia, industry, and state, federal, and international agencies and composed of scientists with infectious disease, animal health, and public health expertise could develop and regularly evaluate and modify a comprehensive, long-term research agenda for animal and zoonotic disease prevention and detection in collaboration with other federal research institutions and laboratories. The advantage of such a group would be to provide a forum for regular exchange of information, sharing of priorities, and possibilities for priority modification. Furthermore, an advisory group could assist federal grant administrators in developing requests for interagency funded proposals for important new areas of animal/zoonotic disease research and emerging animal and zoonotic disease problems. A distinct disadvantage could occur if this group were constructed in such a way as to create another layer of approval or review that would delay the development of new initiatives or implementations of programs.

The comprehensive research agenda could also include an incentive program to encourage academic, government, and the private sectors to partner and to develop novel and effective technologies for the prevention and detection of animal disease both domestically and abroad. As a financial incentive, federal funding agencies could jointly fund new programs (such as program grants by multi-investigator teams from the above sectors) and national centers (integrated zoonotic and wildlife disease research centers and information centers for collecting, collating, and monitoring of diagnostic/disease information across species) to foster and promote collaborative research with the goal to provide an integrated research approach to the detection, diagnosis, and prevention of animal and zoonotic diseases encompassing human and multispecies animal hosts. The advantages of an incentive program would be to incorporate more researchers into the overall strategy to address and control animal diseases and to work and collaborate with other countries on issues of common concern. A disadvantage can arise if a federally funded program cannot exert sufficient control in the appropriate time period to obtain the needed results. The nature of the relationship between federal needs and academic or private industry contributions would have to be structured in such a way that there is continual dialogue and sharing of results in a real-time way.

A more in-depth assessment of national needs for research in animal health is beyond the scope of this report and is being addressed by a forthcoming NRC report on *Critical Needs for Research in Veterinary Science* (NRC, 2005).

Recommendation 5: **To strengthen the animal health and zoonotic disease research infrastructure, the committee recommends that competitive grants be made available to scientists to upgrade equipment for**

animal disease research and that the nation construct and maintain government and university biosafety level 3 (BSL-3 and BSL-3 Ag) facilities for livestock (including large animals), poultry, and wildlife.

Biosafety level 3 facilities (including laboratories and animal rooms for large animals) are needed not only for basic animal disease research, but also for development and evaluation of vaccines and diagnostics for exotic, newly emerging, or highly contagious disease threats. Such facilities will also provide the necessary regional surge capacity for sample and tissue collection and testing from suspect infected animals needed for evaluation of sick or subclinical cases from multiple disease outbreaks. The facilities could in addition be used for BSL-3 vaccine production, storage, or evaluation in emergency outbreak situations.

Prior to September 11, 2001, most animal disease research efforts in the United States were focused on existing indigenous infections of animals and zoonoses and methods for their detection, treatment, prevention, and control. Since then, scientists have begun to focus research efforts on both indigenous and exotic infectious agents of animals and zoonoses in order to develop potential measures against the inadvertent or bioterrorist introduction of potentially devastating diseases of animals, such as foot-and-mouth disease (FMD). Without appropriate countermeasures such as resistant animals, vaccines, or treatments, the U.S. animal and food production systems and related industries are especially vulnerable.

The United States has an extensive system of land-grant universities with scientists having expertise in various arenas of animal diseases and zoonoses. Improving and enhancing the infrastructure of these institutions would be an excellent investment of federal dollars to create national networks of countermeasures for the prevention, detection, and diagnosis of emerging and reemerging animal diseases and zoonoses. The threats of these diseases and bioterror events will continue and likely be accentuated in future years. The December 2003 detection of the first case of BSE in the United States illustrates the continual threat of diseases and their economic impact on both agricultural industries and the national economy, as well as their importance to public health. Enhancement of our current infrastructure for animal disease research is essential to counter the threat of these diseases.

INTERNATIONAL ISSUES

International Interdependence and Collaboration

Recommendation 6: **The United States should commit resources and develop new shared leadership roles with other countries and international organizations in creating global systems for preventing, detect-**

ing, and diagnosing known and emerging diseases, disease agents, and disease threats as they relate to animal and public health.

As the United States and the rest of the world become increasingly interdependent, it is essential to identify animal disease risk factors as they emerge and to focus more attention on the sources and precursors of infections, rather than wait for them to appear in a particular species in the United States. Taken collectively, the recent experience with SARS, West Nile virus, and monkeypox lead to the inescapable conclusion that globalization, population growth, and expansion of human activity into previously unoccupied habitats have essentially connected the United States to potential zoonotic pathogens residing throughout the world. This new reality necessitates coordinated international collaboration directed at identifying potential risks worldwide as well as regulatory mechanisms that minimize the threat of introducing emerging infectious agents into the United States. U.S. support of these efforts, and the cost of implementing them, should be evaluated relative to the potential risks of animal disease to disrupt international markets.

By helping to strengthen other countries' approaches to preventing, detecting, and diagnosing animal diseases, the United States has an opportunity to enhance its own animal health framework. Means to accomplish this include exchange of technology between nations where feasible, and providing training opportunities to international students and veterinarians to ensure self-sufficiency and sustainable surveillance. The United States can also encourage and support the enhancement of critical competencies within the national services, which includes active participation in the formulation of international standards and the timely reporting of zoonotic and exotic diseases. The charge to the committee explicitly states that it will "review the U.S. system and approach for dealing with animal diseases," but the committee regards the international dimension as a critical component of the U.S. animal health framework. As globalization increases—with more movement of diseases, people, products, pathogens, and vectors—the United States cannot continue to impose a line between domestic and international issues.

To operate interdependently means to invest time and resources in building and working through coalitions, formulating international standards that could require years for adoption, and building alliances at the technical level, as well as at the negotiation stages. This is not a short-term activity and requires that the United States make strategic investments to share expertise and better understand and appreciate other countries' current infrastructures, levels of advancements, and greatest challenges and concerns. In determining priority areas, it is important for the United States to work through, support, and promote the leadership of other countries to advance issues and standards of importance in a manner viewed as balanced rather than unilateral. The approach will require a significant investment of time to identify emerging and strategic issues, as well as active outreach and alliances in place to move the plans forward.

Coalition building in international settings will require working more interdependently with other countries. Over the years, the United States has built strong ties with countries with comparable animal health infrastructures, including Canada, Australia, and New Zealand. Similar outreach with emerging countries that are now becoming or will be more active on the international stage should be an equal priority. Proactive approaches—including initiatives undertaken to encourage longer and more sustainable interchanges, sharing of expertise and technical assistance, and pursuing common themes—will build confidence and enable the sharing of data and information at the scientific as well as political levels. This approach is most timely for the early identification of issues associated with the importation, sale, and transport of animals and developing consensus actions to follow.

Importation, Sale, and Transport of Exotic Animals

Recommendation 7: **Integrated and standardized regulations should be developed and implemented nationally to address the import, sale, movement, and health of exotic, non-domesticated, and wild-caught animals.**

Such a policy needs to include health professionals and laboratory-based analysis since many wild-caught and exotic animals may carry pathogens and pose a risk of transmitting disease without demonstrating clinical signs. As noted in Chapter 4, the monkeypox outbreak highlighted weaknesses in the animal health framework for addressing a newly emergent zoonotic disease. In particular, while several federal agencies (including the U.S. Department of Agriculture, the U.S. Department of the Interior's bureau of Fish and Wildlife Service, and the U.S. Department of Health and Human Services) have roles in preventing, detecting, and diagnosing zoonotic and other animal diseases transmitted by exotic animals, there is a lack of coordinated federal oversight of the animal-centered aspects of diseases transmitted by exotic animals. Considering that the emergence of new disease agents occurs most frequently at species interfaces, monkeypox most likely will not be the last zoonotic microbial agent to emerge from an exotic animal in the United States.

The expected beneficial result of this recommendation would be better traceability of exotic and wild companion animals and imported animals. It is essential that regulatory responsibilities for all imported, non-livestock animals be clarified and that the health of these animals be ascertained by appropriate means at the point of origin before importation whenever possible, or at the time of importation when necessary.

ADDRESSING FUTURE ANIMAL DISEASE RISKS

Recommendation 8: **The USDA, DHS, Department of Health and Human Services, and state animal and public health agencies and laboratories should improve, expand, and formalize the use of predictive, risk-based tools and models to develop prevention, detection, diagnostic, and biosecurity systems and strategies for indigenous, exotic, and emerging animal diseases.**

There has been increased recognition and use of well-structured and scientifically based mathematical, epidemiological, and risk analysis models and tools to define acceptable risks and mitigation strategies that can assist in policy and science-based decision making. Examples include models of the spread of FMD during the U.K. epidemic, and an assessment of the risk of BSE to U.S. agriculture, developed by Harvard University's Center for Risk Analysis for the USDA (Cohen et al., 2003; Haydon et al., 2004). Risk analysis and modeling have been criticized, mainly on the basis of insufficient scientific data or inappropriate assumptions. Therefore, efforts to develop scientific data on disease transmission, effectiveness of control programs, economic evaluation, and quantitative assessment of all factors involved in making policies and regulations need to be priorities of the animal health infrastructure as it works in collaboration with academia, industry, and global trade partners.

The World Trade Organization (WTO) Agreement on the Application of Sanitary and Phytosanitary Measures (the SPS Agreement) adopted in 1995 heightens the importance of science- and risk-based tools in evaluating animal disease risks. The SPS Agreement requires governments to adopt sanitary and phytosanitary regulations that facilitate trade in an open, nondiscriminatory, and scientific manner. The concept of risk assessment, which is new for many governments around the world, has risen to the level of an international obligation. Animal health authorities worldwide face a collective challenge in developing risk-based tools and practices that are consistent with their obligations under the SPS Agreement.

Threats from bioterrorism, emerging diseases, and exotic animal disease introductions create an urgent dimension to preventing or minimizing catastrophic consequences to the United States and global economies. Risk-based approaches would proactively address disease threats, identify sources of infection, and respond to predisposing risk factors, including prediction of movement and transmission of disease.

A risk-based approach also calls for committing additional resources to the assessment of risks and consequences of emerging issues. While some of this is occurring, it is currently more driven by the observed economic consequences in other countries and the perceived losses in the United States if a disease occurs than by a fundamental change in mindset toward a risk-based approach. Taking an approach that builds risk profiles and continually monitors and channels intelligence into targeted initiatives or precise actions, long before the presence of disease is detected, is a departure from the past and more in line with what needs to occur in order to allocate resources and efforts effectively. Stakeholders require such information if they are to conduct business differently. Underscoring the risk-based approach is the importance of having professionals who have the education and training to assess, manage, and communicate risk in a manner that supports activities to protect animal health, human health, and the economy. Good communication and information to stakeholders, including producers and the public, are important aspects of an infrastructure that supports risk-based approaches.

EDUCATION AND TRAINING

Recommendation 9: **Industry, producers, the American Veterinary Medical Association (AVMA), government agencies, and colleges of veterinary medicine should build veterinary capacity through both recruitment and preparation of additional veterinary graduates into careers in public health, food systems, biomedical research, diagnostic laboratory investigation, pathology, epidemiology, ecosystem health, and food animal practice.**

This can be achieved through the design and implementation of training and educational curricula to better address these areas and by ensuring that licensing agencies accommodate these new capacities. The Veterinary Workforce Expansion Act of 2005, which amends the Public Health Service Act, will be a useful first step by establishing a competitive grants program to build capacity in veterinary medical education and expand the workforce of veterinarians engaged in public health practice and biomedical research. A critical and fundamental component of this framework is an academic establishment that responds to the nation's needs for educating veterinarians, veterinary specialists, veterinary scientists, and veterinary technologists with the knowledge and skills needed to address emerging zoonoses, changing food animal practices, and other challenges.

As noted in Chapter 4, an inadequate veterinary workforce is a growing problem from the perspective of its distribution, total number, and range of competencies. Contemporary veterinary medicine and that necessary for the future essentially comprises several major and distinct fields dealing with such topics as various species of food-animals, small animals, equine, rural practice (mixed domestic animals), ecosystem health (including wildlife and conservation medicine), public health, and biomedical science. It is equally critical that the human medical and veterinary professions, including regulators, diagnosticians, veterinary practitioners, and owners, routinely have accurate and timely information in order to strengthen early detection capability in the animal health infrastructure.

The committee also recognizes that too few veterinary biomedical scientists and discipline specialists like pathologists are being educated to meet the nation's needs. Forty-three percent of veterinary pathology positions are unfilled, and many pathologists currently working are near retirement (AAVMC, 2004). A previous NRC report, *National Need and Priorities for Veterinarians in Biomedical Research* (NRC, 2004b), has already called attention to the paucity of veterinary researchers in biomedical research and provides a strategy for recruiting and preparing more veterinarians in careers in laboratory animal medicine, comparative medicine, and comparative pathology. This committee endorses the recommendations of that report: to acquaint students with opportunities in comparative medicine throughout veterinary school; increase veterinary school recruitment of applicants with interest or experience in comparative medicine; effect change in veterinary school curricula; address financial barriers to postgraduate training in comparative medicine; increase the number of veterinarians in roles supporting biomedical research; and increase the number of veterinarians serving as principal investigators.

Undergraduate and graduate curricula developed by colleges of veterinary medicine and continuing veterinary education for private practitioners and public (government) service veterinarians must put more emphasis on the contemporary issues in infectious disease epidemiology, risk analysis and management, public health, foreign and emerging diseases, zoonoses, wildlife diseases, bioterrorism, and food safety. In addition, accrediting and licensing agencies and organizations need to be sufficiently flexible to assure that their policies are congruent with this imperative. While increasing veterinary capacity may require more veterinary graduates, the priority is to produce a professional cadre of veterinarians with new skills, knowledge, and abilities that are more responsive to the contemporary and future needs of a changing society. The committee recognizes the desirability of putting forward a comprehensive strategy to increase veterinary capacity, but developing a strategy is beyond the scope of this report.

Recommendation 10: **The USDA, state animal health agencies, the AVMA, and colleges and schools of veterinary medicine and departments of animal science should develop a national animal health education plan focusing on education and training of individuals from all sectors involved in disease prevention and early detection through day-to-day oversight of animals.**

Responsibility for implementing the plan would be at the local level. As noted in Chapter 4, a strong and well-functioning front line of detection is the backbone of effectively controlling animal disease outbreaks.

While different levels of education are required for the various tiers of employees and management, general education and awareness of those who make daily observations of animals should be promoted in order to improve skills in detecting infected or diseased animals. Prerequisite critical education and training, therefore, would include an awareness and recognition of clinical signs, as well as an elementary understanding of disease transmission and prevention. In addition, those with day-to-day oversight of animals need to understand the methods and responsibilities for reporting and the signs of exotic animal diseases. Basic multilingual education and training, with awareness and recognition of biosecurity and implications for breaches in biosecurity, are necessary for those with such direct oversight of animals, whereas managers and owners need more in-depth education to promote greater depth and breadth of understanding of transmission and prevention. Education should be provided formally in English and Spanish, and other languages as necessary, perhaps through mobile education units or long-distance education efforts to minimize time off the job.

IMPROVING PUBLIC AWARENESS OF THE ECONOMIC, SOCIAL, AND HUMAN HEALTH EFFECTS OF ANIMAL DISEASES

Recommendation 11: **The government, private sector, and professional and industry associations should collectively educate and raise the level of awareness of the general public about the importance of public and private investment to strengthen the animal health framework.**

Increased investment in this area will help reduce disease transmission, enhance public and animal health, ensure a secure, economical, and viable food supply, and improve trade and competitiveness. Increased public awareness will be critical in supporting and implementing transformations needed to strengthen the framework against animal disease risks. The lack of a cohesive national advocacy, such as supports public health, creates a much more difficult environment to increase attention and investment in the framework for preventing, detecting, and diagnosing animal diseases. These efforts should include food-animals, wildlife, and companion animals.

In recent years, the news media has given much attention to the unprecedented spread of avian influenza in Asia in January 2004 that killed more than 30 people and led to the slaughter of tens of million of chickens and ducks, the introduction of BSE in the United States, the death of children from *E. coli* O157:H7 after they ate hamburgers at a fast-food restaurant, anthrax-laced letters passing through the U.S. postal service leading to the death of 5 people and infection of another 11, exposure of preschoolers to rabies at a petting zoo in Minnesota, the devastating slaughter of six million British animals in 2001 due to foot-and-mouth disease (Thompson et al., 2002; Haydon et al., 2004), the steady progress of an epidemic of West Nile virus encephalitis in humans and horses across the United States, and the emergence of a new zoonotic disease agent (SARS) in 2003 that sent shock waves around the world, affecting even countries with few cases, like the United States. Despite these news headlines, consumers are largely disconnected from animal agriculture and complacent about the potential costs and risks of animal disease events that lack relevance in their daily lives.

Global disease events in recent years, such as FMD, avian influenza, and BSE, indicate that education and outreach to the consumer, as well as to animal industries, are critical for early detection, for acceptance and compliance with regulatory actions and disease control activities, and for social and economic recovery from catastrophic animal health events. Following detection of BSE in the United Kingdom, contradictory or vague statements from both government offices and the animal agriculture industry often unnerved and confused the general public, ultimately weakening public confidence in policymakers and federal regulators. Similar scenarios have played out with other animal health issues, notably FMD and chronic wasting disease. Therefore, improved information to the public is a critical component of ensuring appropriate levels of public investment in detection, diagnosis, and prevention activities. The nation's framework for addressing animal disease must, of necessity, include a solid foundation in broad-reaching educational programs aimed at improving society's understanding of animal diseases. This applies to all facets of society, including children, consumers, government officials, and health care professionals.

SUMMARY

Extraordinary changes present new threats to animal health, necessitating prompt action within an animal health framework that has not kept pace with science and technology or new global realities. Why is animal health at a critical crossroads? The explosion of human, domestic animal herds, and some wildlife populations—coupled with increased globalization and its industrial development, trade and travel, exploitation of natural resources, and application of modern technology—has greatly increased the degree of contact among humans, domestic animals, and wildlife, and with it, the threat of disease. The growing presence and convergence of just some of these factors calls for a fundamental shift in how regulatory agencies, educators, livestock producers, and industry envision their roles. This report provides a starting point for addressing the needs for improved prevention, diagnosis, and detection. The two planned successor studies (the first one on surveillance and monitoring, the second on response and recovery) will build on the conclusions and recommendations presented in this report. In order to address and to begin building the infrastructure necessary to address critical needs for animal health, a new paradigm for strong leadership, vision, and transformational change will be key in developing dialogue and collaboration among stakeholders. Such collaboration will be important in establishing a mutual understanding that the country's best interest is to be more visionary and strategic and to provide more direct support to efforts that focus on preventing disease rather than only combating disease. This involves:

- Improved development and use of science and technology for prevention and detection.
- Strengthened animal health laboratory networks.
- Comprehensive research with partnered government, academic, and private sectors.
- A coordinating mechanism for engaging partnerships among local, state, federal, and international agencies and the private sector.
- Enhanced global systems for preventing, detecting, and diagnosing diseases.
- A standardized approach for the import, sale, movement, and health of exotic and wild-caught animals.
- Increased use of risk-based tools and models.
- Increased veterinary capacity and capabilities.
- Improved education and training opportunities for individuals responsible for day-to-day oversight of animals.
- Increased awareness about the importance of maintaining animal health.

The evidence discussed in this report provides compelling support for both fundamental changes in the framework related to prevention, detection, and diagnosis of animal diseases and for the urgency in making these changes. The dynamics and realities of today's world require long-term planning and decision making that is well integrated among stakeholders, including international experts and partners. U.S. agencies and stakeholders will have to make significant improvements in their scientific and technological acumen in order to be competitive and to maximize U.S. abilities to sustain and protect animal and public health. The committee is calling for regulatory and oversight agencies to break with the past and engage themselves in this expanded, interdependent role. This will require a large and formal shift away from the introspective mentality of many stakeholder groups and toward a multilateral, open, and transparent operational environment.

References

AAVMC (Association of American Veterinary Medical Colleges). 2003. Emergency Needs in Veterinary Human Resources. Final Report of the AAVMC Task Force, April 15, 2003. Washington, DC: AAVMC.

AAVMC. 2004. Veterinary Medical Education and Workforce Development Act of 2004. September 7, 2004. Washington, DC. Available online at http://aavmc.org/documents/VMEWDA.pdf.

ACVP (American College of Veterinary Pathologists). 2002. Veterinary pathologist survey: final report.

API (Animal Protection Institute). 2003. The dangers of keeping exotic "pets." Available online at http://www.api4animals.org/308.htm.

APPMA (American Pet Products Manufacturing Association). 2005. 2005/2006 National Pet Owners Survey. Greenwich, CT: American Pet Products Manufacturing Association Inc.

AVMA (American Veterinary Medical Association). 2002. U.S. Pet Ownership & Demographics Sourcebook. Available online at http://www.avma.org/membshp/marketstats/sourcebook.asp.

AVMA. 2004a. AVMA News: Changes on the horizon for national veterinary accreditation program. J. Am. Vet. Med. Assoc. 225:819–820.

AVMA. 2004b. Veterinary Market Statistics: First Year Employment. Available online at http://www.avma.org/membshp/marketstats/1yremploy.asp.

AVMA. 2004c. Veterinary Market Statistics: Veterinary Specialists. December 2004. Available online at http://www.avma.org/membshp/marketstats/vetspec.asp.

AVMA. 2005a. AVMA News: USDA takes steps to battle CWD. J Am. Vet. Med Assoc. 226:1962-1963.

AVMA. 2005b. Veterinary Market Statistics: U.S. Veterinarians. Available online at http://www.avma.org/membshp/marketstats/usvets.asp.

Ballesteros, M.L., C.M. Sanchez, and L. Enjuanes. 1997. Two amino acid changes at the N-terminal of the transmissible gastroenteritis coronavirus spike protein result in the loss of enteric tropism. Virology 227:378–388.

Beghin, J., F. Dong, A. Elobeid, J.F. Fabiosa, F.H. Fuller, C.E. Hart, K.P. Kovarik, H. Matthey, A. Saak, and S. Tokgoz. 2004. FAPRI 2004 U.S. and World Agricultural Outlook. Ames, IA: Food and Agricultural Policy Research Institute, Iowa State University and University of Missouri-Columbia. Available online at http://www.fapri.iastate.edu/outlook2004/.

Belay, E.D., R.A. Maddox, E.S. Williams, M.W. Miller, P. Gambetti, and L.B. Schonberger. 2004. Chronic wasting disease and transmission to humans. Emerg. Infect. Dis. 10(6): 977–984.

Benfield, D.A., J.E. Collins, S.A. Dee, P.G. Halbur, H.S. Joo, K.M. Lager, W.L. Mengeling, M.P. Murtaugh, K.D. Rossow, G.W. Stevenson, and J.J Zimmerman. 1999. Porcine reproductive and respiratory syndrome. Pp. 201–232 in Diseases of Swine, 8th Ed., B.E. Straw, S. D'allairte, W.L. Mengeling, and D.J. Taylor, eds. Ames: Iowa State University Press.

Benjamin, G., P.J. Unruh, M.J. Earls, and S.A. Hearne. 2003. Animal-borne Epidemics Out of Control: Threatening the Nation's Health. August 2003. Trust for America's Health. Available online at http://healthyamericans.org/reports/files/Animalreport.pdf.

Bennett, R. 2003. The "direct costs" of livestock disease: the development of a system of models for the analysis of 30 endemic livestock diseases in Great Britain. J. Agri. Econ. 54:55–71.

Bernardo, K., N. Pakulat, M. Macht, O. Krut, H. Seifert, S. Fleer, F. Hunger, and M. Kronke. 2002. Identification and discrimination of *Staphylococcus aureus* strains using matrix-assisted laser desorption/ionization-time of flight mass spectrometry. Proteomics 2:747–53.

Bollinger, T., P. Caley, E. Merrill, F. Messier, M.W. Miller, M.D. Samuel, and E. Vanopdenbosch. 2004. Chronic Wasting Disease in Canadian Wildlife: An Expert Opinion on the Epidemiology and Risks to Wild Deer. Available online at http://wildlife1.usask.ca/ccwhc2003/Publications/CWD%20Expert%20Report%20Final%20-%2020040804.pdf.

Bonner, R.C. 2004. Testimony of Commissioner Robert C. Bonner, U.S. Customs and Border Protection before the National Commission on Terrorist Attacks upon the United States, January 26, 2004. Available online at http://www.customs.ustreas.gov/xp/cgov/newsroom/commissioner/speeches_statements/jan262004.xml.

Bouma, A., A.R. Elbers, A. Dekker, A. deKoeijer, C. Bartels, P. Vellema, P. van der Wal, E.M. van Rooij, F.H. Pluimers, and M.C. de Jong. 2003. The foot-and-mouth disease epidemic in The Netherlands in 2001. Prev. Vet. Med. 57(3):155–166.

Brian, D.A., B.G. Hogue, and T.E. Kienzle. 1995. The coronavirus hemagglutinin esterase clycoprotein. Pp. 165–179 in The Coronaviridae, S.G. Siddell, ed. New York: Plenum Press.

Brown, J.P., and J.D. Silverman. 1999. The current and future market for veterinarians and veterinary medical services in the United States. J. Am. Vet. Med. Assoc. 215:161–183.

Callahan, J.D., F. Brown, F.A. Osorio, J.H. Sur, E. Kramer, G.W. Long, J. Lubroth, S.J. Ellis, K.S. Shoulars, K.L. Gaffney, D.L. Rock, and W.M. Nelson. 2002. Use of a portable real-time reverse transcriptase-polymerase chain reaction assay for rapid detection of foot-and-mouth disease virus. J. Am. Vet. Med. Assoc. 220:1636–1642.

CBC News. 2003. U.S. tentatively links mad cow to Canada. December 27, 2003. Available online at http://www.cbc.ca/stories/2003/12/27/madcow031227.

CBP (U.S. Customs and Border Protection). 2004. Preventing Animal and Plant Pests and Diseases: More than 1.7 million prohibited agricultural items. January 14, 2004. Available online at http://www.customs.gov/xp/cgov/newsroom/press_releases/archives/2004_press_releases/0012004/01142004_4.xml.

CDC (Centers for Disease Control and Prevention). 2003a. Importation of Pets, Other Animals, and Animal Products into the United States. National Center for Infectious Diseases (NCID): Division of Global Migration and Quarantine. Available online at http://www.cdc.gov/ncidod/about.htm.

CDC. 2003b. Multistate outbreak of monkeypox – Illinois, Indiana, and Wisconsin, 2003. Morb. Mort. Wkly. Rep. 52:537–540. Available online at http://www.cdc.gov/mmwr/preview/mmwrhtml/mm5223a1.htm.

CDC. 2003c. Preliminary Report: Multistate Outbreak of Monkeypox in Persons Exposed to Pet Prairie Dogs. CDC National Center for Infectious Diseases: Monkeypox. June 9, 2003. Available online at http://www.cdc.gov/ncidod/monkeypox/pdf/report060903.pdf.

CDC. 2003d. Restrictions on African Rodents and Prairie Dogs, Interim Final Rule. November 4, 2003. Available online at http://edocket.access.gpo.gov/2003/03-27557.htm.

CDC. 2003e. Update: Multistate outbreak of monkeypox – Illinois, Indiana, Kansas, Missouri, Ohio, and Wisconsin, 2003. State and local health departments, Monkeypox investigation team. Morb. Mort. Weekly Rep. 52:642–646.

CDC. 2004a. Emergency Preparedness & Response. Laboratory Preparedness for emergencies. Available online at http://www.bt.cdc.gov/.

CDC. 2004b. Notice of embargo of birds (Class: Aves) from specified Southeast Asian countries. February 4, 2004. Available online at http://www.cdc.gov/flu/avian/pdf/embargo.pdf.

CDC. 2004c. Outbreaks of avian influenza A (H5N1) in Asia and interim recommendations for evaluation and reporting of suspected cases: United States, 2004. Morb. Mort. Wkly. Rep. 53(5):97–100.

CDC. 2004d. West Nile Virus: Transmission Cycle. Vertebrate Ecology. CDC Division of Vector-Borne Infectious Diseases. Available online at http://www.cdc.gov/ncidod/dvbid/westnile/birds&mammals.htm.

CDC. 2004e. West Nile Virus: What you need to know. Available online at http://www.cdc.gov/ncidod/dvbid/westnile/wnv_factsheet.htm.

CDC. 2005. Facts about the Laboratory Response Network. Emergency Preparedness & Response. Available online at http://www.bt.cdc.gov/lrn/pdf/lrnfactsheet.pdf.

Chen, W., M. Yan, L. Yang, B. Ding, B. He, Y. Wang, X. Liu, C. Liu, H. Zhu, B. You, S. Huang, J. Zhang, F. Mu, Z. Xiang, X. Feng, J. Wen, J. Fang, J. Yu, H. Yang, and J. Wang. 2005. SARS-associated coronavirus transmitted from human to pig. Emerg. Inf. Dis. 11:446-448.

Childs, J.E., and G.T. Strickland. 2000. Zoonoses. Pp. 979–984 in Hunter's Tropical Medicine and Emerging Infectious Diseases, 8th ed., G.T. Strickland, ed. Philadelphia, PA: WB Saunders.

Chim, S.S., S.K. Tsui, K.C. Chan, T.C. Au, E.C. Hung, Y.K. Tong, R.W. Chiu, E.K. Ng, P.K. Chan, C.M. Chu, J.J. Sung, J.S. Tam, K.P. Fung, M.M. Waye, C.Y. Lee, K.Y. Yuen, Y.M. Lo; CUHK Molecular SARS Research Group. 2003. Genomic characterisation of the severe acute respiratory syndrome coronavirus of Amoy Gardens outbreak in Hong Kong. Lancet 362(9398):1807–1808.

Chua, K.B. 2003. Nipah virus outbreak in Malaysia. J. Clin. Virol. 26(3):265–275.

CJD Statistics. 2004. The National Creutzfeldt-Jakob Disease Surveillance Unit. Available online at http://www.cjd.ed.ac.uk/figures.htm.

Cleveland, S., M.K. Laurenson, and L.H. Taylor. 2001. Disease of humans and their domestic mammals: pathogen characteristics, host range and the risk of emergence. Phil. Trans. R. Soc. Lond. B. 356: 991–999.

Cohen, J.T., K. Duggar, G.M. Gray, S. Kreindel, H. Abdelrahman, T. HabteMariam, D. Oryang, and B. Tameru. 2003. Revised Risk Assessment: Evaluation of the Potential for Bovine Spongiform Encephalopathy in the United States. Harvard Center for Risk Analysis, Harvard School of Public Health. Prepared for the U.S. Department of Agriculture. November 26, 2001, revised October 2003. Available online at http://www.hcra.harvard.edu/pdf/madcow.pdf

Collinge, J., K.C. Sidle, J. Meads, J. Ironside, and A.F. Hill. 1996. Molecular analysis of prion strain variation and the aetiology of 'new variant' CJD. Nature. 383:666-667.

Colorado State University and Farm Foundation. 2003. The economic impact of animal disease on the animal products sector. Conference on July 11, 2003, in Denver, Colorado. Available online at http://dare.agsci.colostate.edu/animalhealth/conf.htm.

Crespo, R., H.L. Shivaprasad, P.R.Woolcock, R.P. Chin, D. Davidson-York, and R. Tarbell. 1999. Exotic Newcastle disease in a game chicken flock. Avian Dis. 43(2):349–355.

Crosby, A. 1989. America's forgotten pandemic. Cambridge, UK: Cambridge University Press.

Customs and Border Protection Today. 2003. Agriculture's now planted in the field (operations, that is). March 2003. Available online at http://www.customs.gov/xp/CustomsToday/2003/March/ag.xml.

Customs and Border Protection Today. 2004. Introducing the new CBP agriculture specialist. May 2004. Available online at http://www.customs.gov/xp/CustomsToday/2004/May/agSpec.xml.

Daszak, P., A.A. Cunningham, and A.D. Hyatt. 2000. Emerging infectious diseases of wildlife— threats to biodiversity and human health. Science 287:443–449.

Davidson-York, D., R. Tarbell, R.P. Chin, R. Crespo, H.L. Shivaprasad, and P.R. Woolcock. 1998. Velogenic viscerotropic Newcastle disease (VVND) in California. Proceedings of the United States Animal Health Association 102:314–315.

Deem, S.L. 2004. The veterinarian's role in conservation. J. Amer. Vet. Med. Assoc. 225:1033–1034.

DHS (U.S. Department of Homeland Security). 2004a. President's Budget Includes $274 Million to Further Improve Nation's Bio-Surveillance Capabilities. January 29, 2004. Available online at http://www.dhs.gov/dhspublic/display?content=3092.

DHS. 2004b. DHS Organization: Department Components. Available online at http://www.dhs.gov/dhspublic/display?theme=9&content=2973.

DHS. 2004c. Homeland Security Advanced Research Projects Agency (HSARPA): TSWG BIDS, BAA Information Delivery System. Available online at https://www.bids.tswg.gov/hsarpa/bids.nsf/Main?OpenFrameset&67UUR3.

DHS. 2004d. National Response Plan. December 2004. Available online at http://www.dhs.gov/interweb/assetlibrary/NRPbaseplan.pdf.

DHS. 2004e. Protecting Against Agricultural Terrorism. Press Release. February 2, 2004. Available online at http://www.dhs.gov/dhspublic/display?theme=43&content=3117.

Dobson, A., and M. Meagher. 1996. The population dynamics of brucellosis in the Yellowstone National Park. Ecology. 77:1023-1036.

Drosten, C., S. Gunther, W. Preiser, S. van der Werf, H.R. Brodt, S. Becker, H. Rabenau, M. Panning, L. Kolesnikova, R.A. Fouchier, A. Berger, A.M. Burguiere, J. Cinati, M. Eickmann, N. Escriou, K. Grywna, S. Kramma, J.C. Manuguerra, S. Muller, V. Rickerts, M. Sturmer, S. Vieth, H.D. Klenk, A.D. Osterhaus, H. Schmitz, and H.W. Doerr. 2003. Identification of a novel coronavirus in patients with severe acute respiratory syndrome. N. Engl. J. Med. 348:1967–1976.

Dubey J.P., and D.S. Lindsay. 1996. A review of *Neospora caninum* and neosporosis. Vet. Parasitol. 67:1–59.

EAAP (European Association for Animal Production). 2003. After BSE—a future for the European livestock sector. E.P. Cunningham and the European Association for Animal Production, eds. Available online at http://www.wageningenacademic.com/books/EAAP108.pdf.

Easterday, B.C., V.S. Hinshaw, and D.A. Halvorson. 1997. Influenza. In Diseases of Poultry, 10th edition, B.W. Calneck, ed. Ames: Iowa State University Press.

Eisner, R. 1991. Veterinary researchers: let more money go to the dogs. The Scientist 5(1):10. Available online at http://www.the-scientist.com/1991/1/7/10/1.

Erles, K., C. Toomey, H.W. Brooks, and J. Brownlie. 2003. Detection of a novel canine coronavirus in dogs with respiratory disease. Abstract 4.2 in IXth Intl. Symp. on Nidoviruses, The Netherlands, May 24–29, 2003.

Eyre, P. 2002. Engineering veterinary education. J. Vet. Med. Educ. 29(4):195–200.

FAIR (Food Animal Integrated Research). 2002. FAIR 2002: Animal products for the next millennium, an agenda for research and education Available online at http://www.fass.org/fair2002.pdf.

FAO (Food and Agriculture Organization) and WHO (World Health Organization). 2001. Codex Alimentarius Commission: Procedural Manual, Twelfth Edition. Joint FAO/WHO Food Standards Programme. Available online at http://www.fao.org/documents/show_cdr.asp?url_file=/DOCREP/005/Y2200E/y2200e07.htm.

Fenner, F. 1970. The effects of changing social organisation on the infectious diseases of man. Pp. 48–68 in The Impact of Civilization on the Biology of Man, S.V. Boyden, ed. Canberra: Aust. Natl. Univ. Press.

Fouchier, R.A., T. Kuiken, M. Schutten, G. van Amerongen, G.J. van Doornum, G.B. van den Hoogen, M. Peiris, W. Lim, K. Stohr, and A.D. Osterhaus. 2003. Aetiology: Koch's postulates fulfilled for SARS virus. Nature 423:240.

Frazier, M.E., G.M. Johnson, D.G. Thomassen, C.E. Oliver, and A. Patrinos. 2003. Realizing the potential of the genome revolution: the genomes to life program. Science 300:290–293.

FSIS (USDA Food Safety and Inspection Service). 2003. Backgrounders/Key Facts: Risk Analysis, July 2003. US Department of Agriculture, Washington, DC. Available online at http://www.fsis.usda.gov/oa/background/riskanal.htm.

FWS (U.S. DOI Fish and Wildlife Service). 2001. U.S. Fish & Wildlife Service Division of Law Enforcement, Annual Report FY 2001. Available online at http://library.fws.gov/Pubs9/LEannual01.pdf.

Gamblin, S.J., L.F. Haire, R.J. Russell, D.J. Stevens, B. Xiao, Y. Ha, N. Vasisht, D.A. Steinhauer, R.S. Daniels, A. Elliot, D.C. Wiley, and J.J. Skehel. 2004. The structure and receptor binding properties of the 1918 influenza hemagglutinin. Science 303:1838–1842.

GAO (U.S. Government Accountability Office). 2000. Agricultural research: USDA's response to recommendations to strengthen the Agricultural Research Service's programs and facilities. GAO/RCED-OO-85R ARS Programs and Facilities. Letter to John Kasich, Chairman, Committee on the Budget, House of Representatives. February 15.

Gilchrist, M. 2001. The progress, priorities and concerns of public health laboratories. Institute of Medicine: Forum on Emerging Infections, Biological Threats and Terrorism, November 28, 2001. Washington, DC. Available online at http://www.nap.edu/openbook.php?record_id=10290&page=160.

Glass, W.G., K. Subbarao, B. Murphy, and P.M. Murphy. 2004. Mechanisms of host defense following severe acute respiratory syndrome-coronavirus (SARS-CoV) pulmonary infection of mice. J. Immunol. 173:4030-4039.

Gross J.E., and M.W. Miller. 2001. Chronic wasting disease in mule deer: disease dynamics and control. J. Wildlife Mgt. 65:205–215.

Guan, Y., B.J. Zheng, Y.Q. He, X.L. Liu, Z.X. Zhuang, C.L. Cheung, S.W. Luo, P.H. Li, L.J. Zhang, Y.J. Guan, K.M. Butt, K.L. Wong, K.W. Chan, W. Lim, K.F. Shortridge, K.Y. Yuen, J.S. Peiris, and L.L. Poon. 2003. Isolation and characterization of viruses related to the SARS coronavirus from animals in southern China. Science 302:276–278.

Hamels, S., J.L. Gala, S. Dufour, P. Vannuffel, N. Zammatteo, and J. Remacle. 2001. Consensus PCR and microarray for diagnosis of the genus *Staphylococcus*, species, and methicillin resistance. Biotechniques 31:1364–1366, 1368, 1370–1372.

Hampton, T. 2004. Clues to the deadly 1918 flu revealed. JAMA 291(13):1553–1554.

Haydon D.T., R.R. Kao, and P.K. Kitching. 2004. The UK foot and mouth disease outbreak – the aftermath. Nature Reviews 2:675–681.

Hennessy, D.A., J. Roosen, and H.H. Jensen. 2005. Infectious disease, productivity, and scale in open and closed animal production systems. Am. J. Agr. Econ. (in press, November 2005).

Herrewegh, A.A.P.M., M. Mähler, H.J. Hedrich, B.L. Haagmans, H.F. Egerink, M.C. Horzinek, P.J.M. Rottier, and R.J. de Groot. 1997. Persistence and evolution of feline coronavirus in a closed cat-breeding colony. Virol. 234:349–363.

HHS (U.S. Department of Health and Human Services). 1999. Biosafety in Microbiological and Biomedical Laboratories, 4th Edition. Washington, DC: U.S. Government Printing Office. Available online at http://www.cdc.gov/od/ohs/pdffiles/4th%20BMBL.pdf.

HHS. 2004. Statement from the Department of Health and Human Services Regarding Chiron Flu Vaccine. HHS Press Office, October 5, 2004. Available online at http://www.hhs.gov/news/press/2004pres/20041005.html.

Hibler C.P., K.L. Wilson, T.R. Spraker, M.W. Miller, R.R. Zink, L.L. DeBuse, E. Andersen, D. Schweitzer, J.A. Kennedy, L.A. Baeten, J.F. Smeltzer, M.D. Salman, and B.E. Powers. 2003. Field validation and assessment of an enzyme-linked immunosorbent assay for detecting chronic wasting disease in mule deer (*Odocoileus virginianus*), and Rocky Mountain elk (*Cervus elaphus nelsoni*). J. Vet. Diagn. Invest. 15:311–319.

Hird, D., L. King, M. Salman, and R. Werge. 2002. A crisis of lost opportunity: conclusions from a symposium on challenges for animal population health education. J. Vet. Med. Educ. 29:205–209.

Hoblet, K.H., A.T. Maccabe, and L.E. Heider. 2003. Veterinarians in population health and public practice: meeting critical national needs. J. Vet. Med. Educ. 30:287–294.

Huang, S.H., T. Triche, and A.Y. Jong. 2002. Infectomics: genomics and proteomics of microbial infections. Funct. Integr. Genomics 1:331–344.

Humanitarian Resource Institute. 2004. CDC List of Select Agents: USDA List of High Consequence Livestock Diseases: OIE List A-B. Foreign Animal and Zoonotic Disease Center. July 30, 2004. Available online at http://www.humanitarian.net/biodefense/fazdc/fadr30704.html.

Hunter, N., J. Foster, A. Chong, S. McCutcheon, D. Parnham, S. Eaton, C. MacKenzie, and F. Houston. 2002. Transmission of prion diseases by blood transfusion. J. Gen.Virol. 83:2897–2905.

Hutin Y.J.F., R.J. Williams, P. Malfait, R. Pebody, V.N. Loparev, S.L. Ropp, M. Rodriguez, J.C. Knight, F.K. Tshioko, A.S. Khan, M.V. Szczeniowski, and J.J. Esposito. 2001. Outbreak of human monkeypox, Democratic Republic of Congo, 1996 to 1997. Emerg. Infect. Dis. 7:434–438.

IAEA (International Atomic Energy Agency). 2001. The Agency's Programme and Budget for 2001. Available online at http://www.iaea.org/About/Policy/GC/GC44/Documents/gc44-6.pdf#xml=http://www.iaea.org/search97cgi/.

IFPRI (International Food Policy Research Institute). 1999. World Food Prospects: critical issues for the early twenty-first century. Available online at http://www.ifpri.org/pubs/fpr/fpr29.pdf.

IFPRI. 2003. IFPRI's strategy toward food and nutrition security. Available online at http://www.ifpri.org/about/gi14.pdf.

IOM (Institute of Medicine). 1998. Ensuring Safe Food from Production to Consumption. Washington, DC: National Academy Press.

IOM. 2003. Microbial Threats to Health: Emergence, Detection, and Response. Washington, DC: The National Academies Press.

IOM. 2005. The Threat of Pandemic Influenza: Are We Ready? Workshop Summary. Washington, DC: The National Academies Press.

Ismail, M.M., K.O. Cho, L.A. Ward, L.J. Saif, and Y.M. Saif. 2001. Experimental bovine coronavirus in turkey poults and young chickens. Avian Dis. 45(1):157–163.

Jackson, R.J., A.J. Ramsay, C.D. Christensen, S. Beaton, D.F. Hall, and I.A. Ramshaw. 2001. Expression of mouse interleukin-4 by a recombinant ectromelia virus suppresses cytolytic lymphocyte responses and overcomes genetic resistance to mousepox. J. Virol. 75:1205–1210.

JAVMA News. 2004. Feline leukemia virus threatens endangered panthers. J. Am. Vet. Med. Assoc. 224:1721–1722.

Karg, M. 2000. Designated licensure: the case for specialization within the veterinary degree. J. Am. Vet. Med. Assoc. 217:1792–1796.

Khodakevich, L., Z. Jezek, and D. Kinzana. 1986. Isolation of monkeypox from a wild squirrel. Lancet 1:98–99.

Kirkwood, J.K., and A.A. Cunningham. 1994. Epidemiological observations on spongiform encephalopathies in captive wild animals in the British Isles. Vet. Rec. 135:296-303.

Ksiazek, T.G., D. Erdman, C.S. Goldsmith, S.R. Zaki, T. Peret, S. Emery, S. Tong, C. Urbani, J. A. Comer, W. Lim, P.E. Rollin, S.F. Dowell, A.E. Ling, C.D. Humphrey, W.J. Shieh, J. Guarner, C.D. Paddock, P. Rota, B. Fields, J. DeRisi, J.Y. Yang, N. Cox, J.M. Hughes, J. W. LeDuc, W.J. Bellini, and L.J. Anderson. 2003. A novel coronavirus associated with severe acute respiratory syndrome. N. Engl. J. Med. 348:1953–1966.

Kuchler, F., and S. Hamm. 2000. Animal disease incidence and indemnity eradication programs. Agr. Econ. 22:299–308.

Kuiken, T., G. Rimmelzwaan, D. van Riel, G. van Amerongen, M. Baars, R. Fouchier, and A. Osterhaus. 2004. Avian H5N1 influenza in cats. Science. 306:241.

Landyl, I.D., P. Ziegler, and A. Kima. 1972. A human infection caused by monkeypox virus in Basankusu Territory, Democratic Republic of the Congo (DRC). Bull. World Health Organ. 46:593–597.

Langkop, C.W., C. Austin, M. Dworkin, K. Kelly, H. Messersmith, R. Teclaw, J. Howell, J. Michael, P. Pontones, G. Pezzino, G.R. Hansen, M.V. Wegner, J.J. Kazmierczak, C. Williams, D.R. Croft, S. Ahrabi-Fard, L. Will, H.H. Bostrom, J.P. Davis, A.Fleischauer, M. Sotir, G. Huhn, R. Kanwal, J. Kile, and J. Sejvar. 2003. Update: Multistate outbreak of monkeypox – Illinois, Indiana, Kansas, Missouri, Ohio, and Wisconsin, 2003. June 20, 2003. Morb. Mort. Weekly Rep. 52:561–564.

Laude, H., K. Van Reeth, and M. Pensaert. 1993. Porcine respiratory coronavirus: molecular features and virus-host interactions. Vet. Res. 24:125–150.

Lee, J.W., and W.J. McKibbin. 2004. Chapter 2: Political influences on the response to SARS and economic impacts of the disease: estimating the global economic costs of SARS. Pp. 92–109 in Learning from SARS: Preparing for the Next Disease Outbreak – Workshop Summary. Institute of Medicine. Washington, DC: National Academies Press.

Leighton, F.A. 2003. West Nile Virus. Available online at http://wildlife1.usask.ca/ccwhc2003/wildlife_health_topics/arbovirus/arbown.htm.

Ludwig, G.V., P.P Calle, J.A. Mangiafico, B.L. Raphael, D.K. Danner, J.A. Hile, T.L. Clippinger, J.F. Smith, R.A. Cook, and T. McNamara. 2002. An outbreak of West Nile virus in a New York City captive wildlife population. Am. J. Trop. Med. Hyg. 67(1):67–75.

Mansley, L.M., P.J. Dunlop, S.M. Whiteside, and R.G. Smith. 2003. Early dissemination of foot and mouth disease virus through sheep marketing in February 2001. Vet. Rec. 153:43–50.

Martina, B.E.E., B.L. Haagmans, T. Kuiken, R.A.M. Fouchier, G.F. Rimmelzwaan, G. van Amerongen, J.S.M. Peiris, W. Lim, and A.D.M.E. Osterhaus. 2003. Virology: SARS virus infection of cats and ferrets. Nature 425:915.

McAuliffe, J., L. Vogel, A. Roberts, G. Fahle, S. Fischer, W.J. Shieh, E. Butler, S. Zaki, M. St.Claire, B. Murphy, and K. Subbarao. 2004. Replication of SARS coronavirus administered into the respiratory tract of African green, rhesus, and cynomolgus monkeys. Virology. 330:8-15.

McElroy, R., J. Johnson, M. Morehart, J. Ryan, C. McGath, R. Green, A. Mishra, J. Hopkins, T. Covey, K. Erickson, and W. McBride. 2003. Agricultural Income and Finance Outlook, US Department of Agriculture, Economic Research Service, AIS-81, November 5, 2003. Washington, DC Available online at http://usda.mannlib.cornell.edu/reports/erssor/economics/ais-bb/2003/ais81.pdf.

McElwain, T.F. 2003. Agriculture Outlook Forum 2003: Establishing an animal disease diagnostic network. February 20, 2003. Available online at www.usda.gov/oce/waob/Archives/2003/speeches/McElwain%20B&W.doc.

Meteyer, C.U., D.E. Docherty, L.C. Glaser, J.C. Franson, D.A. Senne, and R. Duncan. 1997, Diagnostic findings in the 1992 epornitic of neurotropic velogenic Newcastle disease in double-crested cormorants from the upper midwestern United States. Avian Dis. 41:171-180.

Miller, M.W., E.S. Williams, C.W. McCarty, T.R. Spraker, T.J. Kreeger, C.T. Larsen, and E.T. Thorne. 2000. Epizootiology of chronic wasting disease in free-ranging cervids in Colorado and Wyoming. J.Wildlife Dis. 36:676–690.

Miller, M.W., E.S. Williams, N.T. Hobbs, and L. L. Wolfe. 2004. Environmental sources of prion transfer in mule deer. Emerg. Infect. Dis. 10:1003–1006.

Molenda, S.L. 2003. History of Newcastle disease in U.S. Available online at http://www.internationalparrotletsociety.org/historyofend.html.

Moore, D.A. 2003. Introducing the theme–continuing veterinary medical education: where are we? J. Vet. Med. Educ. 30(1):12.

Moore, D.A., D.J. Klingborg, and T. Wright. 2003. Mandatory continuing veterinary medical education requirements in the United States and Canada. J. Vet. Med. Educ. 30(1):19–27.

NASDARF (National Association of State Departments of Agriculture Research Foundation). 2001. The Animal Health Safeguarding Review: Results and Recommendations. Available online at http://www.aphis.usda.gov/vs/pdf_files/safeguarding.pdf

National Audit Office. 2002. The 2001 outbreak of foot and mouth disease: report by the comptroller and auditor general. HC 939 Session 2001–2002. 21 June 2002. London: Stationery Office. Available online at http://www.nao.org.uk/publications/nao_reports/01-02/0102939.pdf.

Nielsen, N.O. 2003. Will the veterinary profession flourish in the future? J. Am. Vet. Med. Educ. 30:301–307.

Nolen, R.S. 1999. Veterinarians key to discovering outbreak of exotic encephalitis. J. Am. Vet. Med. Assoc. 215:1415,1418–1419.

Nolen, R.S. 2002. Exotic Newcastle disease strikes game birds in California. J. Am. Vet. Med. Assoc. 221(10):1369–1370.

Normile, D. 2004.Viral DNA match spurs China's civet cat roundup. Science 303:292.

NRC (National Research Council).1972. New Horizons for Veterinary Medicine. Washington, DC: National Academy Press.

NRC. 1982. Specialized Veterinary Manpower Needs Through 1990. Washington DC: National Academy Press.

NRC. 1994a. Livestock Disease Eradication: Evaluation of the Cooperative State-Federal Bovine Tuberculosis Eradication Program. Washington, DC: National Academy Press.

NRC. 1994b. Science and Judgment in Risk Assessment. Washington, DC: National Academy Press.

NRC. 1998. Brucellosis in the Greater Yellowstone Area. Washington, DC: National Academy Press.

NRC. 2002. Emerging Animal Diseases: Global Markets, Global Safety: Workshop Summary. Washington, DC: The National Academies Press.

NRC. 2003a. Countering Agricultural Bioterrorism. Washington, DC: The National Academies Press.

NRC. 2003b. Diagnosis and Control of Johne's disease. Washington, DC: The National Academies Press.

NRC. 2003c. Frontiers in Agriculture Research: Food, Health, Environment, and Communities. Washington, DC: The National Academies Press.

NRC. 2004a. Advancing Prion Science: Guidance for the National Prion Research Program. Washington, DC: The National Academies Press.

NRC. 2004b. National Need and Priorities for Veterinarians in Biomedical Research. Washington, DC: The National Academies Press.

NRC. 2004c. Science and Judgment in Risk Assessment. Washington, DC: The National Academies Press.

NRC. 2004d. Science, Medicine, and Animals. Washington, DC: The National Academies Press.

NRC. 2005. Critical Needs for Research in Veterinary Science. Washington, DC: The National Academies Press.

Ogden, H.G. 1987. CDC and the Smallpox Crusade; U.S. Department of Health and Human Services; HHS publication No. (CDC) 87–8400. Washington, DC: U.S. Government Printing Office, pp. 73–74.

OIE (Office International des Epizooties). 2003. Terrestrial Animal Health Code Section 1.3.2: Guidelines for Import Risk Analysis. Available online at http://www.oie.int/eng/normes/mcode/en_chapitre_1.3.2.htm.

OIRSA (Organismo Internacional Regional de Sanidad Agropercuaria). 2004. Objetivo y Marco de Referencia. Available online at http://www.oirsa.org/DTSA/Objetivo.htm.

O'Rourke, K.I., T.V. Baszler, S.M. Parish, and D.P. Knowles. 1998. Preclinical detection of PrPSc in nictitating membrane lymphoid tissue of sheep. Vet. Rec. 18:489–491.

O'Rourke, K.I., T.V. Baszler, T.E. Besser, J.M. Miller, R.C. Cutlip, G.A. Wells, S.J. Ryder, S.M. Parish, A.N. Hamir, N.E. Cockett, A. Jenny, and D.P. Knowles. 2000. Preclinical diagnosis of scrapie by immunohistochemistry of third eyelid lymphoid tissue. J. Clin. Microbiol. 9:3254–3259.

OSTP (The White House, Office of Science and Technology Policy). 2005. Science and technology: A foundation for homeland security. April 2005. Available online at http://www.ostp.gov/html/OSTPHomeland.pdf.

Otte, M.J., R. Nugent, and A. McLeod. 2004. Transboundary Animal Diseases: Assessment of Socio-Economic Impacts and Institutional Responses. Livestock Discussion Paper No. 9. Food and Agriculture Organization, Livestock Information and Policy Branch, AGAL. February 2004.

Peckenpaugh, J. 2003. No Retraining for Agricultural Inspectors in Border Agency Plan. October 29, 2003. Available online at http://www.govexec.com/dailyfed/1003/102903p1.htm.

Peiris, J.S., S.T. Lai, L.L. Poon, Y. Guan, L.Y. Yam, W. Lim, J. Nicholls, W.K. Yee, W.W. Yan, M.T. Cheung, V.C. Cheng, K.H. Chan, D.N. Tsang, R.W. Yung, T.K. Ng, and K.Y. Yuen. 2003. Coronavirus as a possible cause of severe acute respiratory syndrome. Lancet. 361:1319–1325.

Peiris, J.S., Y. Guan, and K.Y. Yuen. 2004. Severe acute respiratory syndrome. Nature Medicine. 10:S88-S97.

Pensaert, M.B. 1999. Porcine epidemic diarrhea. Pp. 179–185 in Diseases of Swine, 8th Ed., B.E. Straw, S. D'allairte, W.L. Mengeling, and D.J. Taylor, eds. Ames: Iowa State University Press.

Poutanen, S.M., D.E. Low, B. Henry, S. Finkelstein, D. Rose, K. Green, R. Tellier, R. Draker, D. Adachi, M. Ayers, A.K. Chan, D.M. Skowronski, I. Salit, A.E. Simor, A.S. Slutsky, P.W. Doyle, M. Krajden, M. Petric, R.C. Brunham, and A.J. McGeer. 2003. Identification of severe acute respiratory syndrome in Canada. N. Engl. J. Med. 348:1995–2005.

Presley, J. 2004. Mad cow to cost firms almost $6 billion. Organic Consumers Association. February 25, 2004. Available online at http://www.organicconsumers.org/madcow/billion225-4.cfm.

Preston, R. 1997. The Cobra Event. New York: Random House.

Pritchard, W.R. 1988. Future Directions for Veterinary Medicine. Durham, NC: Pew National Veterinary Education Program, Duke University.

Radostits, O.M. 2003. Engineering veterinary education: a clarion call for reform in veterinary education. J. Vet. Med. Educ. 30:176–190.

Ribble C., B. Hunter, N. Lariviere, D. Belanger, G. Wobeser, P.Y. Daoust, T. Leighton, D. Waltner-Toews, J. Davidson, E. Spangler, and O. Nielsen. 1997. Ecosystem health as a clinical rotation for senior students in Canadian veterinary schools. Can. Vet. J. 38(8):485–490.

Riddle C., H. Mainzer, and M. Julian 2004. Training the veterinary public health workforce: a review of educational opportunities in U.S. veterinary schools. J. Vet. Med. Educ. 31:161–167.

Roberts, A., L. Vogel, J. Guarner, N. Hayes, B. Murphy, S. Zaki, and K. Subbarao. 2005. Severe acute respiratory syndrome coronavirus infection of golden Syrian hamsters. J. Virol. 79:503-511.

Rossides, S.C. 2002. A farming perspective on the 2001 foot and mouth disease epidemic in the United Kingdom. Rev. Sci. Tech. Off. Int. Epiz. 21(3):831–838.

Royal Society. 2002. Royal Society Inquiry commissioned by the UK government into infectious diseases in livestock. Available online at www.royalsoc.ac.uk/inquiry/index.html.

Saif, L.J. 2004. Comparative biology of coronaviruses: lessons for SARS. In SARS: The First New Plague of the 21st Century, M. Peiris, ed. Oxford, UK: Blackwell Pub.

Saif, L., and R. Wesley. 1999. Transmissible gastroenteritis virus. In Diseases of Swine, 8th Ed., B.E. Straw, S. D'allairte, W.L. Mengeling, and D.J. Taylor, eds. Ames: Iowa State University Press.

Sanchez, C.M., A. Izeta, J.M. Sanchez-Morgado, S. Alonso, I. Sola, M. Balasch, J. Plana-Duran, and L. Enjuanes. 1999. Targeted recombination demonstrates that the spike gene of transmissible gastroenteritis coronavirus is a determinant of its enteric tropism and virulence. J. Virol. 73:7607–7618.

Saritelli, R.A. 2001. Point of view: from Machiavelli to mad cow disease: 20th century ecological changes and the inevitable role of medicine in disease prevention. Med. Health R I. 84(11): 379–381.

Shadduck, J.A., R. Storts, and L.G. Adams. 1996. Selected examples of emerging and re-emerging infectious diseases in animals. Am. Soc. of Microbiol. News 62:586–588.

Shepherd, A.J. 2004. Employment, starting salaries, and educational indebtedness of year-2004 graduates of US veterinary medical colleges. J. Am. Vet. Med. Assoc. 224:1677–1679.

Stabel, J.R. 1996. Production of interferon-alpha by peripheral blood mononuclear cells: an important diagnostic tool for detection of subclinical paratuberculosis. J. Vet. Diagn. Invest. 8:345–350.

STDF (Standards and Trade Development Facility). 2004. Standards and Trade Development Facility. Available online at http://www.standardsfacility.org.

Stevens, J., A.L. Corper, C.F. Basler, J.K. Taubenberger, P. Palese, and I.A. Wilson. 2004. Structure of the uncleaved human H1 hemagglutinin from the extinct 1918 influenza virus. Science 303:1866–1870.

Sumner, D.A. 2003. Economics of policy for exotic pests and diseases: principles and issues. In Exotic Pests and Diseases: Biology and Economics for Biosecurity, D. Sumner, ed. Ames: Iowa State University Press.

Swanson, M. 2004. Wells Fargo Economics: Monthly Agriculture Market Update Report. November 2004. Available online at http://a248.e.akamai.net/7/248/1856/2be47190b08305/www.wellsfargo.com/downloads/pdf/com/research/agricultural/agupnov04crops.pdf.

Thompson, D., P. Muriel, D. Russell, P. Osborne, A. Bromley, M. Rowland, S. Creigh-Tyte, and C. Brown. 2002. Economic costs of the foot and mouth disease outbreak in the United Kingdom in 2001. Rev. Sci. Tech. 21(3):675–687.

Thorne, E.T., and E.S. Williams. 1988. Disease and endangered species: the blackfooted ferret as a recent example. Conserv. Biol. 2(1):66–74.

Thurmond, M., and C. Brown. 2002. Bio- and agroterror: the role of the veterinary academy. J. Vet. Med. Educ. 29:1–4.

Tomassen, F.H., A. deKoeijer, M.C. Mouritis, A. Dekker, A. Bouma, and R.B. Huirne. 2002. A decision-tree to optimise control measures during the early stage of a foot-and-mouth disease epidemic. Prev. Vet. Med. 54(4):301–324.

Torres A., and Q.P. Bowman. 2002. New Directions for the national veterinary accreditation program. J. Am. Vet. Med. Assoc. 220:1470–1472.

Torrey, E.F. and R.H. Yolken. 2005. Beasts of the Earth. New Brunswick, NJ: Rutgers University Press, p. 12.

Tsunemitsu, H., H. Reed, Z. El-Kanawati, D. Smith, and L.J. Saif. 1995. Isolation of coronaviruses antigenically indistinguishable from bovine coronavirus from a Sambar and white tailed deer and waterbuck with diarrhea. J. Clin. Microbiol. 33:3264–3269.

Ungchusak K., P. Auewarakul, S.F. Dowell, R. Kitphati, W. Auwanit, P. Puthavathana, M. Uiprasertkul, K. Boonnak, C. Pittayawonganon, N.J. Cox, S.R. Zaki, P. Thawatsupha, M. Chittaganpitch, R. Khontong, J.M. Simmerman, and S. Chunsutthiwat. 2005. Probable person-to-person transmission of avian influenza A (H5N1). N. Engl. J. Med. 352(4):333–340.

University of Georgia. 2003. Velogenic Newcastle Disease, USDA Foreign Animal Diseases. The Gray Book. Available online at http://www.aphis.usda.gov/lpa/pubs/fsheet_faq_notice/fs_ahend.html.

USAHA (United States Animal Health Association). 2002. Report of the Joint USAHA/AAVLD Committee on Animal Health Information Systems. Available online at http://www.usaha.org/reports/report02/r02ahis.html.

USAHA. 2003a. Newsletter: Special Issue. October 2003. Available online at http://www.usaha.org/news/newsletter/USAHA-Newsletter-Oct2003.pdf.

USAHA. 2003b. Report of the USAHA Committee on Wildlife Diseases. Chairman: Dr. Michael W. Miller, Ft. Collins, CO; Vice Chairman: Dr. John R. Fischer, Athens, GA. Available online at http://usaha.org/committees/reports/reports03/r03wd.html.

USAHA. 2003c. Resolution No. 21 from the 2003 USAHA Annual Meeting Oct. 15, 2003 Available online at http://www.usaha.org/resolutions/reso03/res-2103.html.

USAHA. 2005. USAHA: The Nation's Animal Health Forum Since 1897. Available online at http://www.usaha.org/about_usaha.shtml.

USAID (U.S. Agency for International Development). 1992. USAID Policy Determination: Definition of Food Security, PD-19. April 13, 1992. Available online at http://www.usaid.gov/policy/ads/200/pd19.pdf.

U.S. Congress. 1914. Smith-Lever Act of 1914. 7 USC. 341 et seq., 38 Stat. 372. Available online at http://higher-ed.org/resources/smith.htm.

U.S. Congress. 1994. P.L. (Public Law) 103–354. Federal Crop Insurance Reform and Department of Agriculture Reorganization Act of 1994.

USDA (U.S. Department of Agriculture). 2001. Data submitted to the National Research Council Committee on Opportunities in Agriculture from the Agricultural Research Service. Washington, DC: Agricultural Research Service, U.S. Department of Agriculture.

USDA. 2002a. Agriculture Fact Book: 2001–2002. Washington, DC: U.S. Government Printing Office. Available online at http://www.usda.gov/factbook/2002factbook.pdf.

USDA. 2002b. ARS Facilities Design Standards. AFM/ARS Facilities Engineering Branch. July 24, 2002. Available online at http://www.afm.ars.usda.gov/ppweb/242-01m.htm.

USDA. 2002c. 2002 Census of Agriculture. Available online at http://www.nass.usda.gov/census/.

USDA. 2002d. Research, Education, and Economics Strategic Plan. Washington, DC: U.S. Department of Agriculture. Available online at http://www.reeusda.gov/ree/ree2.htm.

USDA. 2002e. Veterinary Biologics: Uses and Regulation. Program Aid No. 1713. APHIS, March 2002. Available online at http://www.aphis.usda.gov/lpa/pubs/pub_ahvetbiologic.html

USDA. 2002f. USDA Releases $43.5 Million to States for Strengthening Agriculture Homeland Security Protections. Available online at http://www.usda.gov/news/releases/2002/05/0213.htm.

USDA. 2003. Food Market Structures: the U.S. food and fiber system. October 16, 2003. Available online at http://ers.usda.gov/briefing/foodmarketstructures/foodandfiber.htm.

USDA. 2004a. National Aquatic Animal Health Plan. USDA-APHIS-Veterinary Services. Available online at http://www.aphis.usda.gov/vs/aqua/naah_plan.html.

USDA. 2004b. National Veterinary Services Laboratories. USDA-APHIS-VS. Available online at http://www.aphis.usda.gov/vs/nvsl/index.htm. 2004.

USDA. 2004c. Spring Viremia of Carp, United States. USDA-APHIS-VS Center for Emerging Issues. July 20, 2004, Impact Statement. Available online at http://www.aphis.usda.gov/vs/ceah/cei/IW_2004_files/svc_mo_070804_files/SVCMO070804final.htm.

USDA. 2004d. Veterinary Services Strategic Plan FY 2004 to FY 2008. Updated February 2004. Available online at http://www.aphis.usda.gov/vs/pdf_files/strat_plan.pdf.

USDA. 2004e. West Nile Virus Veterinary Services Factsheet. USDA-APHIS. October 2004. Available online at http://www.aphis.usda.gov/lpa/pubs/fsheet_faq_notice/fs_ahwnv.html.

USDA. 2005. Fact Sheet: Bovine Spongiform Encephalopathy (BSE) EPI report. June 2005. Available online at http://www.usda.gov/documents/FactSheetbse062905.pdf.

USDA APHIS (Animal and Plant Health Inspection Service). 2002. Plan for Assisting States, Federal Agencies, and Tribes in Managing Chronic Wasting Disease in Wild and Captive Cervids (June 26, 2002). Available online at http://www.aphis.usda.gov/lpa/issues/cwd/cwd62602.html.

USDA APHIS. 2003a. The Animal and Plant Health Inspection Service and Department of Homeland Security: Working Together to Protect Agriculture. APHIS Fact Sheet. May 2003. Available online at http://www.aphis.usda.gov/lpa/pubs/fsheet_faq_notice/fs_aphis_homeland.html.

USDA APHIS. 2003b. Exotic Newcastle Disease (END) – United States Update: Summary of Selected Events, January – June 2003. APHIS Veterinary Services, Centers for Epidemiology and Animal Health, Center for Emerging Issues. Available online at http://www.aphis.usda.gov/vs/ceah/cei/IW_2003_files/SummaryDisease_2003.htm.

USDA APHIS. 2003c. Summary of Selected Disease Events, January – June 2003. Center for Emerging Issues. Available online at http://www.aphis.usda.gov/vs/ceah/cei/DiseaseSummary_files/summary_1_to_6_2003_files/disease_summary010603.htm.

USDA APHIS. 2004a. Emergency Management Warning 105: Exotic Newcastle Disease in the United States. Available online at www.aphis.usda.gov/lpa/issues/enc/emws/encdw105-03.html.

USDA APHIS. 2004b. West Nile virus: Equine program information. Available online at http://www.aphis.usda.gov/vs/nahps/equine/wnv/maps.html.

USDA APHIS-VS (Veterinary Services). 1991. Qualitative analysis of BSE risk factors in the United States. Fort Collins, CO: Centers for Epidemiology and Animal Health.

USDA APHIS-VS. 1997. Virus-Serum-Toxin Act (as amended December 23, 1985). Available online at http://www.aphis.usda.gov/vs/cvb/vsta.htm.

USDA APHIS-VS. 2001. Part I: Reference of swine health and management in the United States, 2000. August 2001. Fort Collins, CO: National Animal Health Monitoring System (NAHMS). Also available online at http://www.aphis.usda.gov/vs/ceah/ncahs/nahms/swine/swine2000/Swine2kPt1.pdf.

USDA APHIS-VS. 2002. Infectious Salmon Anemia. APHIS Fact Sheet. January 2002. Available online at http://www.aphis.usda.gov/lpa/pubs/iaisa.html.

USDA APHIS-VS. 2004. National Veterinary Accreditation Program (NVAP): About NVAP. August 31, 2004. Available online at http://www.aphis.usda.gov/vs/nvap/index.html.

USDA APHIS-VS. 2005. Chronic Wasting Disease. Available online at http://www.aphis.usda.gov/vs/nahps/cwd/.

USDA ARS (Agricultural Research Service). 2002. Development and Validation of Rapid Diagnostic Tests for Avian Influenza and Newcastle Disease. Available online at http://www.ars.usda.gov/research/projects/projects.htm?accn_no=405127.

USDA ARS. 2004. Research: National Programs. Available online at http://www.ars.usda.gov/research/programs.htm.

USDA ERS (Economic Research Service). 2001. Agricultural Outlook. August 2001. pp. 4–6. Available online at http://usda.mannlib.cornell.edu/reports/erssor/economics/ao-bb/2001/ao283.pdf.

USDA ERS. 2004. US Agricultural Trade Update. USDA-ERS. Available online at http://usda.mannlib.cornell.edu/reports/erssor/trade/fau-bb/.

USGS (U.S. Geological Survey). 2004. National Wildlife Health Center. Available online at http://www.nwhc.usgs.gov.

Utterback, W. 1973. Epidemiology of VVND in Southern California. Proceedings of the United States Animal Health Association 76: 280–287.

Van Leeuwen, J.A., N.O. Nielsen, and D. Waltner-Toews. 1998. Ecosystem health: an essential field for veterinary medicine. J. Am. Vet. Med. Assoc. 212:53–57.

Waldner C.L., E.D. Janzen, J. Henderson, and D. M. Haines 1999. An outbreak of abortion associated with *Neospora caninum* infection in a beef herd. J. Amer. Vet. Med. Assoc. 215:1485–1490.

Walsh, D.A., F.A. Murphy, B.I. Osburn, L. King, and A.M. Kelly. 2003. Executive Summary. An agenda for action: Veterinary medicine's crucial role in public health and biodefense and obligation of academic veterinary medicine to respond. J. Vet. Med. Educ. 30(2): 92–95.

Watts, J. 2004. China culls wild animals to prevent new SARS threat. Lancet 363:134.

White House, The. 2004. Homeland Security Presidential Directive/HSPD-9, January 30, 2004. Available online at http://www.whitehouse.gov/news/releases/2004/02/20040203-2.html.

Whitener, L.A., and D.A. McGranahan. 2003. Rural America: opportunities and challenges. Amber Waves. Washington, DC: USDA Economic Research Service.

WHO (World Health Organization). 2003. Avian Influenza in the Netherlands. April 24, 2003. WHO: Communicable Disease Surveillance and Response (CSR). Available online at http://www.who.int/csr/don/2003_04_24/en/.

WHO. 2004 Avian Influenza Fact Sheet. 15 January 2004. Available online at http://www.who.int/csr/don/2004_01_15/en.

WHO. 2005. Cumulative Number of Confirmed Human Cases of Avian Influenza A(H5N1) since 28 January 2004. Available online at http://www.who.int/csr/disease/avian_influenza/country/cases_table_2005_01_07/en/.

Williams E.S., and S. Young. 1980. Chronic wasting disease of captive mule deer: a spongiform encephalopathy. J. Wildlife Dis. 16:89–98.

Williams, E.S., M.W. Miller, and E.T. Thorne. 2002. Chronic wasting disease: implications and challenges for wildlife managers. Transactions of the North American Wildlife and Natural Resources Conference 67. Available at http://www.cwd-info.org/pdf/CWD PresentationatNAWNRC.pdf.

Wines, M. 2004. For sniffing out land mines, a platoon of twitching noses. New York Times, May 18, 2004.

Wise, J.K., and A.J. Shepherd. 2004. Employment of male and female graduates of US veterinary medical colleges. 2003. J. Amer. Vet. Med. Assoc. 224(4):517–519.

WTO (World Trade Organization). 1998. Understanding the WTO Agreement on Sanitary and Phytosanitary Measures, May 1998. Available online at http://www.wto.org/english/tratop_e/sps_e/spund_e.htm.

Wu, D., C. Tu, C. Xin, H. Xuan, Q. Meng, Y. Liu, Y. Yu, Y. Guan, Y. Jiang, X. Yin, G. Crameri, M. Wang, C. Li, S. Liu, M. Liao. L. Feng, H. Xiang, J. Sun, J. Chen, Y. Sun, S. Gu, N. Liu, D. Fu, B.T. Eaton, L.F. Wang, and X. Kong. 2005. Civets are equally susceptible to experimental infection by two different Severe Acute Respiratory Syndrome coronavirus isolates. J. Virol. 79:2620-2625.

Xu, R., J.F. He, M. R. Evans, G.W. Peng, H.E. Fields, D.W. Yu, C.K. Lee, H.M. Luo, W.S. Lin, P. Lin, L.H. Li, W.J. Liang, J.Y. Lin, and A. Schnur. 2004. Epidemiologic clues to SARS origin in China. Emerg. Infect. Dis. 10:1030-1037.

Zhang, X.M., W. Herbst, K.G. Kousoulas, and J. Storz. 1994. Biologic and genetic characterization of a hemagglutinating coronavirus isolated from a diarrhoeic child. J. Med. Virol. 44:152–161.

Zheng, B.J., Y.Guan, K. H. Wong, J. Zhou, K. L. Wong, B.W.Y. Young, L.W. Lu and S.S. Lee. 2004. SARS-related virus predating SARS outbreak, Hong Kong. Emerg. Inf. Dis. 10:176–178.

APPENDIXES

Appendix A

Acronyms and Abbreviations

AAVLD	American Association of Veterinary Laboratory Diagnosticians
AAVMC	Association of American Veterinary Medical Colleges
ABSL	Animal biosafety levels
AFMIC	Armed Forces Medical Intelligence Center (DoD)
AI	Avian influenza
AIQ	Agricultural inspection and quarantine
APHIS	Animal and Plant Health Inspection Service (USDA)
ARS	Agricultural Research Service
AVMA	American Veterinary Medical Association
BANR	Board on Agriculture and Natural Resources
BIA	Bureau of Indian Affairs
BLM	Bureau of Land Management
BSE	Bovine spongiform encephalopathy, also mad cow disease
BSL	Biosafety level
BTS	Border and Transportation Security (DHS)
CBP	Bureau of Customs and Border Protection (DHS-BTS)
CDC	Centers for Disease Control and Prevention
CEAH	Center for Epidemiology and Animal Health (USDA-APHIS-VS)
CFSAN	Center for Food Safety and Applied Nutrition (FDA)
COE	Council on Education (AUMA)
CoV	Coronavirus

CSF	Classical swine fever
CSREES	Cooperative State Research, Education, and Extension Service (USDA)
CVB	Center for Veterinary Biologics (USDA-APHIS-VS)
CVM	Center for Veterinary Medicine (FDA)
CWD	Chronic wasting disease
DHS	U.S. Department of Homeland Security
DOC	U.S. Department of Commerce
DoD	U.S. Department of Defense
DOI	U.S. Department of the Interior
DOS	U.S. Department of State
DVM	Doctor of Veterinary Medicine
EID	Emerging infectious diseases
EP	Emergency Programs
EPA	U.S. Environmental Protection Agency
EM	Electron microscopy
EMS	Emergency medical services
END	Exotic Newcastle disease
FAD	Foreign animal disease
FADDL	Foreign Animal Disease Diagnostic Laboratory
FAO	Food and Agriculture Organization
FAS	Foreign Agricultural Service (USDA)
FBI	Federal Bureau of Investigation
FDA	Food and Drug Administration
FEMA	Federal Emergency Management Agency
FMD	Foot-and-mouth disease
FMDV	Foot-and-mouth disease virus
FSIS	Food Safety and Inspection Service (USDA)
FWS	Fish and Wildlife Service (DOI)
GOARN	Global Alert and Response Network
HACCP	Hazard analysis and critical control point
HHS	U.S. Department of Health and Human Services
HSPD	Homeland Security Presidential Directive
HS-Centers	University-Based Homeland Security Centers of Excellence
IAIP	Informational Analysis and Infrastructure Protection Directorate

IOM	Institute of Medicine (National Academy of Sciences)
IS	International Services (USDA-APHIS)
LRN	Laboratory Response Network
NAHLN	National Animal Health Laboratory Network
NAVLE	North American Veterinary Licensing Examination
NBII	National Biological Information Infrastructure (USGS)
NCID	National Center for Infectious Diseases (CDC)
NCIE	National Center for Import and Export (USDA-APHIS-VS)
NIH	National Institutes of Health
NMFS	National Marine Fisheries Service (DOC-NOAA)
NOAA	National Oceanic and Atmospheric Administration
NPS	National Park Service
NSF	National Science Foundation
NVAP	National Veterinary Accreditation Program (USDA-APHIS-VS)
NVSL	National Veterinary Services Laboratories (USDA-APHIS-VS)
NWHC	National Wildlife Health Center (USGS)
OIE	Office International des Epizooties, also World Organization for Animal Health
OIG	Office of the Inspector General
PCR	Polymerase chain reaction
PEDV	Porcine epidemic diarrhea virus
PIADC	Plum Island Animal Disease Center (DHS)
POE	Port(s) of entry
PPQ	Plant Protection and Quarantine (USDA-APHIS)
PRRSV	Porcine respiratory and reproductive syndrome virus
RT-PCR	Reverse transcriptase-polymerase chain reaction
SARS	Severe acute respiratory syndrome
SECWDS	Southeastern Cooperative Wildlife Disease Study
S&T	Science and Technology Directorate (DHS)
TGEV	Transmissible gastroenteritis virus
TME	Transmissible mink encephalopathy
TSE	Transmissible spongiform encephalopathy
U.K.	United Kingdom
USAHA	U.S. Animal Health Association

USAID	U.S. Agency for International Development
USAMRIID	U.S. Army Medical Research Institute for Infectious Diseases
USDA	U.S. Department of Agriculture
USGS	U.S. Geological Survey
USTR	U.S. Trade Representative
VMD	Veterinariae Medicinae Doctor (Doctor of Veterinary Medicine)
VPH	Veterinary Public Health (WHO)
VS	Veterinary Services (USDA-APHIS)
WHO	World Health Organization
WNV	West Nile virus
WS	Wildlife Services (USDA-APHIS)
WTO	World Trade Organization

Appendix B

Glossary of Terms

Animal biosafety level 3 (ABSL-3) facilities: Involves practices suitable for work with animals infected with indigenous or exotic agents that present the potential of aerosol transmission and of causing serious or potentially lethal disease. ABSL-3 builds upon the standard practices, procedures, containment equipment, and facility requirements of ABSL-2.

Animal health framework: The collection of organizations and participants in the public and private sectors who are directly responsible for maintaining the health of all animals who are impacted by animal disease or influence its determinants.

Antiviral: Destroying or inhibiting the growth and reproduction of viruses.

Assay: Qualitative or quantitative analysis of a substance. In the context here, assay refers to determining presence of a toxin, chemical, infectious agent, or antibodies to an agent.

Avian influenza (AI): A disease of viral etiology that ranges from a mild or even asymptomatic infection to an acute, fatal disease of chickens, turkeys, guinea fowls, and other avian species, especially migratory waterfowl (1,2,3,4,8,9,10,11).

Biosafety level: Laboratories are assigned a classification (levels 1 to 4) based on the risk to human health of handling certain types of organisms.

Level 1 laboratories are designed for low-risk work; level 4 laboratories can handle organisms that pose the most serious risks. Laboratories at each classification level must meet different design criteria and conform to different operating procedures. The University of Georgia AHRC building will house level 2 and 3 laboratories.

Biosafety level 1 (BSL-1) is used for working with agents having no known or minimal hazard to laboratory personnel and the environment; the organisms are unlikely to cause illness in people or animals.

- Work is generally conducted on open bench tops with standard microbiological practices.
- Examples: *Bacillus subtilis*, nonpathogenic *E. coli*

Biosafety level 2 (BSL-2) is suitable for work involving agents of moderate potential hazard to personnel and the environment. Should a person become infected, treatment is available, and the risk of spreading the infection to others is low.

- Any laboratory procedure with these agents that may create an aerosol must be done within a biological safety cabinet.
- Examples: *Salmonella* spp., *Shigella* spp., most animal viruses

Biosafety level 3 (BSL-3) is applicable to work done with agents that may cause serious illness to people or animals but cannot spread easily to others; treatment is available.

- *All* procedures involving the manipulation of infectious materials must be done within biological safety cabinets or other appropriate containment devices. The laboratory has special engineering and design features including separation from traffic flow, water-resistant surfaces for cleaning, sealed windows, and ducted exhaust air ventilation.
- Examples: virulent Newcastle disease virus, HIV research level, *Coxiella burnettii* (Q fever), *E. coli* 0157:H7

Bioterror: A form of terrorism that employs the use of biological and chemical weapons.

Bovine tuberculosis: Tuberculosis in cattle caused by infection with the bacterium *Mycobacterium bovis*, which can be transmitted to other animals and to humans.

Brucellosis: A disease of domestic animals, such as cattle, sheep, goats, and dogs, that is caused by brucellae and sometimes results in spontaneous abortions in newly infected animals. In humans it is caused by any of several species of *Brucella* and marked by fever, sweating, weakness, and headache. It is transmitted to humans by direct contact with diseased animals or through ingestion of infested meat, milk, or cheese.

Cervid: Any of various hoofed ruminant mammals of the family Cervidae, characteristically having deciduous antlers borne chiefly by the males. The deer family also includes the elk, moose, caribou, and reindeer.

Classical swine fever (CSF): A highly contagious, deadly disease of swine, also known as hog cholera.

Coronavirus: Any of various single-stranded, RNA-containing viruses that cause respiratory infection in humans and resemble a crown when viewed under an electron microscope because of their petal-shaped projections.

Chronic wasting disease (CWD): A transmissible spongiform encephalopathy (TSE) affecting elk and deer (cervids) in North America.

Endemic: The pattern of disease characterized by a sustained level of disease over time.

Epidemic: The pattern of disease characterized by an increase in frequency of disease above the expected for the population.

Exotic animal disease: Diseases such as SARS and monkeypox that are new and/or emerging diseases, but which are not listed by OIE. In the report, the term refers to any animal disease caused by a disease agent that does not naturally occur in the United States.

Exotic Newcastle disease (END): A contagious and fatal viral disease affecting all species of birds.

Foot-and-mouth disease (FMD): A highly contagious viral infection primarily of cloven-hoofed domestic animals (cattle, pigs, sheep, goats, and water buffalo) and cloven-hoofed wild animals. The disease is characterized by fever and vesicles with subsequent erosions in the mouth, nares, muzzle, feet, or teats.

Foreign animal disease (FAD): Long-standing diseases that have been kept out of the United States (e.g., FMD, CSF, BSE, rinderpest, etc.) and that are listed by OIE (list A and list B). In the report, the term refers to an exotic animal disease limited to agricultural animals.

Hog cholera: A highly contagious viral disease of swine that occurs in an acute, subacute, chronic, or persistent form. In the acute form, the disease is characterized by high fever, severe depression, multiple superficial and internal hemorrhages, and high morbidity and mortality. In the chronic form, the signs of depression, anorexia, and fever are less severe than in the acute form, and recovery is occasionally seen in mature animals.

Infectious: Capable of causing infection; communicable by invasion of the body of a susceptible organism.

Johne's disease: A chronic inflammatory bowel disease, primarily in cattle, caused by *Mycobacterium paratuberculosis.*

Monkeypox: A rare viral disease that is found mostly in the rain forest countries of central and western Africa. The disease is called "monkeypox" because it was first discovered in laboratory monkeys in 1958.

Necropsy: An examination and dissection of a dead body to determine cause of death or the changes produced by disease.

Pandemic: An epidemic that occurs worldwide.

Pathogen: Disease-producing organism.

Polymerase chain reaction: A technique for amplifying DNA sequences in vitro by separating the DNA into two strands and incubating it with oligonucleotide primers and DNA polymerase. It can amplify a specific sequence of DNA by as many as one billion times and is important in biotechnology, forensics, medicine, and genetic research.

Prion: A microscopic protein particle similar to a virus but lacking nucleic acid, thought to be the infectious agent responsible for scrapie and certain other degenerative diseases of the nervous system.

Protozoa: Any of a large group of single-celled, usually microscopic, eukaryotic organisms, such as amoebas, ciliates, flagellates, and sporozoans.

Pseudorabies: A highly contagious herpes virus infection of animals (especially pigs) that affects the central nervous system.

Q fever: A disease that is characterized by high fever, chills, muscular pains, headache, and sometimes pneumonia, that is caused by a rickettsial bacterium of the genus *Coxiella* (*C. burnetii*) of which domestic animals serve as reservoirs, and that is transmitted to humans especially by inhalation of infective airborne bacteria (as in contaminated dust).

Rinderpest: An acute, often fatal, contagious viral disease, chiefly of cattle, characterized by ulceration of the alimentary tract and resulting in diarrhea.

SARS (Severe acute respiratory syndrome): A viral respiratory illness caused by a coronavirus, called SARS-associated coronavirus (SARS-CoV).

Serology: The science that deals with the properties and reactions of serums, especially blood serum, and typically relates to the testing of sera for antibodies against viruses or bacteria.

Surveillance: An active, systematic, ongoing, and formal process aimed at early detection of a disease, an agent, or elevated risk of disease in a population.

Toxin: A poisonous substance produced during the metabolism and growth of certain microorganisms and some higher plant and animal species.

Transmissible spongiform encephalopathy (TSE): Examples include, but are not limited to, the following diseases: feline spongiform encephalopathy, bovine spongiform encephalopathy, chronic wasting disease, and scrapie. See individual species.

Vaccine: Substance administered to animal to stimulate its defense mechanism.

West Nile virus (WNV): The mosquito-borne virus that causes West Nile fever, one of the flaviviruses, a family of viruses also responsible for dengue, yellow fever, and tick-borne encephalitis virus; like the other flaviviruses, WNV is a positive-strand RNA virus containing three structural proteins and a host-derived lipid bilayer.

Zoonoses: Diseases caused by infectious agents that can be transmitted between (*or are shared by*) animals and humans.

Appendix C

Existing Federal System for Addressing Animal Diseases

A White Paper

Prepared By
Nga L. Tran, Dr.PH, MPH, CIH

of

Exponent

TABLE OF CONTENTS

Service (ARS)
1.2.4.3 DHS-Science and Technology (S&T) Directorate
1.2.4.4 DHHS - NIH
1.2.4.5 DOD—US Army Medical Research, Institute of Infectious Diseases (USAMRIID)
1.2.5 Emergency Response and Communication
1.2.5.1 Existing Federal Emergency Response Plan
1.2.5.2 Federal Response to a FMD outbreak or similarly infectious diseases
1.2.5.3 Communication
1.3 BUDGET AND CAPACITY
1.3.1 Laboratory Capacity
1.3.2 Veterinarian Capacity

2 DISEASES IN WILDLIFE

2.1 The Department of the Interior (DOI), US Fish and Wildlife Service (FWS)
2.2 The Department of Interior, Bureau of U.S. Geological Survey (USGS), Biological Resources Division (BRD), the National Wildlife Health Center (NWHC)
2.3 The Department of Interior, USGS, National Biological Information Infrastructure (NBII) Programs
2.3.1 The NBII Wildlife Disease Information Node (WDIN)
2.4 USDA-APHIS-VS' Wildlife Service (WS)

3 FISHERIES

3.1 Department of Commerce, National Oceanic and Atmospheric Administration (NOAA), National Marine Fisheries Service (NMFS)

4 FOOD SAFETY

4.1 USDA-FSIS
4.2 FDA
4.3 CDC

ACRONYMS AND ABBREVIATIONS

AAVLD	American Association of Veterinary Laboratory Diagnosticians
AFMIC	DoD-Armed Forces Medical Intelligence Center
AIQ	Agricultural inspection and quarantine
APHIS	USDA Animal and Plant Health Inspection Service
AEOC	APHIS Emergency Operations Center
ARS	Agricultural Research Service
AVIC	Area veterinarian in charge
BSE	Bovine spongiform encephalopathy, also mad cow disease
BSL	Biosafety level
BTS	Border and Transportation Security (DHS)
CBP	Bureau of Customs and Border Protection (DHS-BTS)
CADIA	Center for Animal Disease Information and Analysis (CEAH)
CDC	Centers for Disease Control and Prevention
CEAH	Center for Epidemiology and Animal Health (USDA-APHIS-VS)
CFSAN	Center for Food Safety and Applied Nutrition (FDA)
CIA	Central Intelligence Agency
CSREES	USDA- Cooperative State Research, Education, and Extension Service
CVB	Center for Veterinary Biologics (APHIS-VS)
CVM	Center for Veterinary Medicine (FDA)

DHS	U.S. Department of Homeland Security
DHHS	U.S. Department of Health and Human Services
DOC	U.S. Department of Commerce
DOE	U.S. Department of Energy
DoD	U.S. Department of Defense
DOI	U.S. Department of Interior
DOJ	U.S. Department of Justice
DOS	U.S. Department of States
DOT	U.S. Department of Transportation
EPA	U.S. Environmental Protection Agency
EMS	Emergency Management Response
FAD	Foreign animal diseases
FADD	Foreign animal disease diagnostician
FADDL	Foreign Animal Disease Diagnostic Laboratory
FAS	Foreign Agricultural Service (USDA)
FBI	Federal Bureau of Investigation
FDA	Food and Drug Administration
FEMA	Federal Emergency Management Agency
FMD	Foot-and-mouth disease
FSIS	Food Safety and Inspection Service
FWS	U.S. Fish and Wildlife Service
HHS	U.S. Department of Health and Human Services
HSPD	Homeland Security Presidential Directive
HS-Centers	University-Based Homeland Security Centers of Excellence
IAHI	International animal health information
INS	Immigration and Naturalization Service (DOJ)
IS	International Services (USDA-APHIS)
JSA	Joint Subcommittee on Aquaculture
LRN	Laboratory Response Network
MOU	Memorandum of understanding
NAAHC	North American Animal Health Committee
NADC	National Animal Disease Center (USDA-ARS)
NAHEMS	The National Animal Health Emergency Management System

NAHLN	National Animal Health Laboratory Network
NAHMS	National Animal Health Monitoring System
NAHRS	National Animal Health Reporting System
NBII	National Biological Information Infrastructure (USGS)
NBACC	National Biodefense Analysis Countermeasure Center (DHS-S&T)
NCIE	National Center for Import and Export (USDA-APHIS-VS)
NCID	National Center for Infectious Diseases (CDC)
NMFS	National Marine Fisheries Service (DOC-NOAA)
NIAA	National Institute of Animal Agriculture
NIC	National Incidence Coordinator (USDA)
NIMS	National Incident Management System
NOAA	National Oceanic and Atmospheric Administration
NRMT	USDA-APHIS-National Response Management Team
NRP	National Response Plan
NSS	National Surveillance System (USDA-APHIS-VS)
NSU	National Surveillance Unit (USDA-APHIS-VS)
NVSL	National Veterinary Services Laboratories (USDA-APHIS-VS)
NWHC	National Wildlife Health Center (USGS)
NWRC	National Wildlife Research Center (APHIS-VS)
OIE	Office International des Epizooties
OIG	Office of Inspector General
PIADC	Plum Island Animal Disease Center (DHS)
POE	Port(s) of entry
PPQ	Plant Protection and Quarantine (USDA-APHIS)
READEO	Regional Emergency Animal Disease Eradication Organization (APHIS-VS)
S&T	Science and Technology Directorate (DHS)
SITC	Smuggling Interdiction and Trade Compliance
SCO	State Coordinating Officer
TECS	Treasury Enforcement Communications System (U.S. Treasury)
USAHA	U.S. Animal Health Association
USAIP	U.S. Animal Identification Plan
USARMRIID	U.S. Medical Research Institute for Infectious Disease

USDA	U.S. Department of Agriculture
USERPS	U.S. Emergency Response Plan System
USFWS	U.S. Fish and Wildlife Service
USGS	U.S. Geological Survey
USTR	U.S. Trade Representative
VMO	Veterinary medical officer
VS	Veterinary Services (USDA-APHIS)
WDIN	Wildlife Disease Information Node (USGS)
WS	Wildlife Services (USDA-APHIS)

EXISTING FEDERAL SYSTEM FOR ADDRESSING ANIMAL DISEASES

During the past 10 years, emerging and reemerging pathogens have become a major human and animal health concern. The globalization of trade, increased international travel, changing weather patterns, rapid population growth in cities, intensive agriculture, limited genetic diversity in farm animals, and changes in farm practices are creating new opportunities for the reemergence and spread of infectious diseases in both humans and livestock.[1] Reservoirs of infection in the wild also pose constant and increasing threats to domestic livestock population.[2] Cost of losses from disease in livestock and poultry in the United States is currently estimated at over $17.5 billion dollars per year.[3] Further, nearly 200 zoonotic diseases can be naturally transmitted from animals to man.[4]

Rapid diagnostic tests, novel genetic vaccines, vigilance in monitoring and surveillance, and increased biosecurity measures will be needed to effectively detect and control emerging diseases and to prevent future animal disease outbreaks. Expanded research will also be needed to accelerate the development of information and technologies for the protection of U.S. livestock, poultry, wildlife and human health against zoonotic diseases. A complex infrastructure for protecting animal health has arisen at the federal level from a number of statutory mandates and regulatory authorities. This infrastructure is based on a large number of programmatic components of several federal agencies. A flow diagram of the existing organizational web of the major agencies involved in efforts to ensure animal health in the United States is presented in Figure C-1.

This paper outlines thc existing legal authorities that establish the existing federal infrastructure for addressing animal diseases. Program-

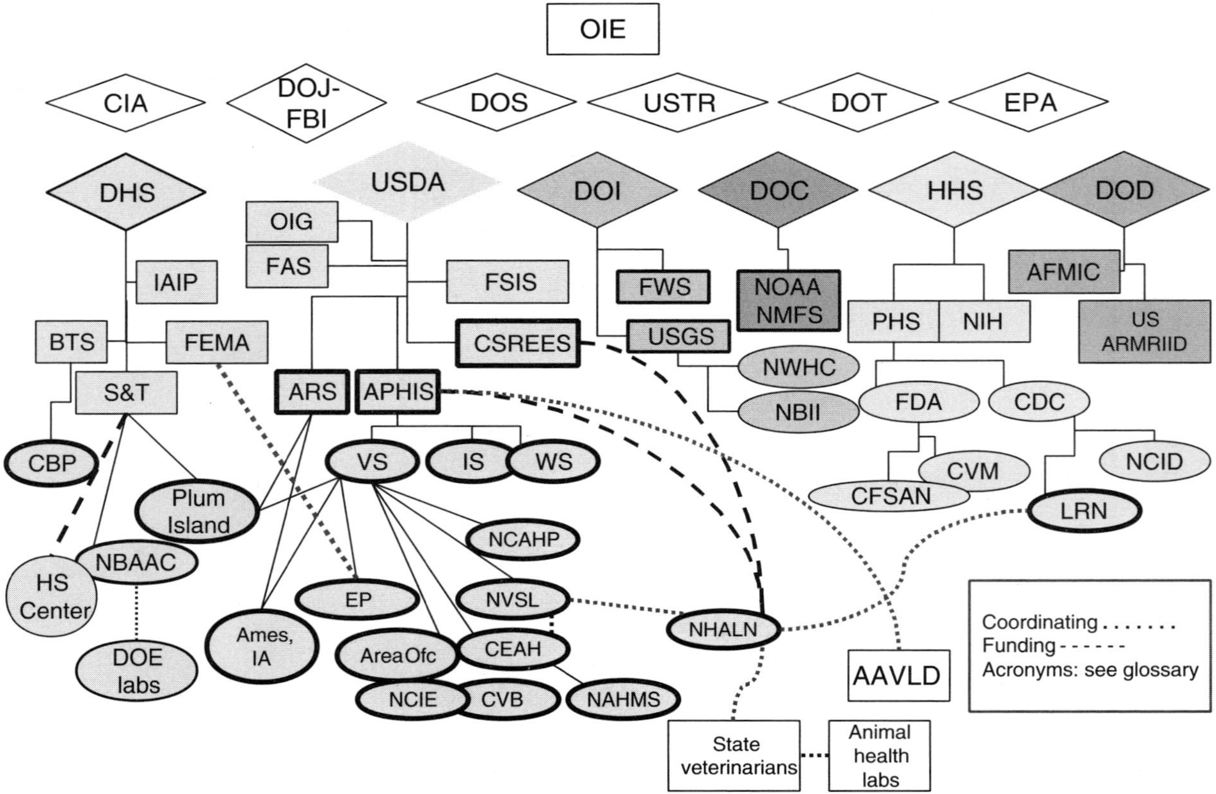

FIGURE C-1 Existing federal system for addressing animal diseases: An organizational web.

matic activities to carry out the following functions: deterrence and prevention, monitoring and surveillance, detection and diagnosis, research and education, and emergency response and communication, are profiled in details for key federal agencies that are directly involved in the management of farm animal health, disease prevention and monitoring, and response to disease outbreaks. Readily available budget and capacity information is also summarized for several key USDA agencies and DHS. Because of the potential for wildlife to impact farm animals, the programmatic functions performed by federal agencies involved in the management of wildlife diseases are also profiled. Since fisheries are part of the U.S. agricultural system, federal agencies responsible for their health management are also described, albeit in a limited form, in this report. Finally, diseases in farm animals could have direct impact food safety; thus, federal agencies managing food safety are briefly summarized.

1 FARM ANIMALS

1.1 LEGAL FRAMEWORK

The USDA Animal and Plant Health Inspection Service (APHIS) has the lead role in protecting animal health. The Animal Health Protection Act passed into law on May 13, 2002, (Public Law 107-171; Farm Security and Rural Investment Act of 2002, 116 Stat. 134) repealing previous Animal Health and Quarantine laws and providing the Secretary of USDA broad authority and discretion to prevent, detect, control, and eradicate diseases and pests of animals.[5] Animal disease means any infectious or noninfectious disease or condition affecting the health of livestock or any condition detrimental to production or marketing of livestock.[6] The Secretary of USDA has delegated the responsibility to APHIS.

Within APHIS, the majority of the responsibility to protect animal health resides in Veterinary Services (VS). VS also derives its authorities from the Virus-Serum-Toxin Act of 1913. Recently, the Agricultural Bioterrorism Protection Act added responsibilities for overseeing agents or toxins deemed a severe threat to animal health.[7,8] Other USDA agencies, including the USDA Agricultural Research Services (ARS) and the USDA Cooperative State Research, Education, and Extension Service (CSREES), also play critical roles in protecting animal health. The legal and regulatory foundation for APHIS and other USDA agencies that engage in activities to address agricultural animal diseases are summarized in Table C-1.

Interstate and Foreign Quarantine regulations (42 CFR70 and 71) authorize the Secretary of the Department of Health and Human Services (DHHS), through the Centers for Disease Control and Prevention (CDC) to make and enforce regulations to prevent transmission of infectious disease from foreign countries into the United States. Under these authorities, CDC can set policy to embargo certain animals from entering the United States.[9] Title III of the Bioterrorism Act provides the Secretary of the Department of Health and Human Services with new authorities to protect the nation's food supply. DHHS legal authorities in addressing animal diseases are summarized in Table C-1.

The Public Health Security and Bioterrorism Preparedness Response Act requires notification and controls on the movement of agents or toxins deemed to be a threat to animal or plant health and to animal and plant products. To prevent the incursions of adverse animal health events, USDA-APHIS units are working with DHHS to implement the provisions of this act.[23]

The Homeland Security Act of 2002 establishes the Department of Homeland Security (DHS) and its directorates. More than 22 federal agencies were consolidated into the new department, including components of the USDA-APHIS that conduct inspection and animal quarantine activities at U.S. ports and Plum Island Animal Disease Center (PIADC).[24] Table C-2 provides a description of statutory authorities for the various DHS directorates with Homeland Security programmatic responsibilities that are either directly or indirectly addressing animal diseases.

1.2 FUNCTIONS

As authorized by various laws and regulations, the USDA agencies' programs that address agricultural animal health cover a wide range of functions. In this paper, these functions are grouped as follows: deterrence and prevention, monitoring and surveillance, detection and diagnosis, emergency response, research, education and training, and communication. The programmatic activities that address these functions by various USDA agencies are summarized in Table C-3. Other federal agencies carry out a number of these functions to protect agricultural animal health. In the following sections, existing programs at USDA and other federal agencies performing these functions are described in more detail.

1.2.1 Deterrence and Prevention

Deterrence and prevention are considered the first lines of defense against the introduction of animal and plant pests and pathogens from foreign or domestic sources.[35] Several strategies are involved including global and regional strategies that are directed at reducing a potential threat before it reaches the U.S. borders, and border strategy that focuses on interdicting a threat agent at U.S. ports of entry.[36]

TABLE C-1 Legal Authorities

Agency	Legal and Regulatory Foundation	Authorities
USDA, Animal and Plant Health Inspection Service (APHIS)	• The Animal Industry Act of 1884 as amended (21USC 117) • The Cattle Contagious Diseases The Act of 1903 as amended (21 USC 111-115, 117, 120, 123, 125-127, 134) • The Farm Security and Rural Investment Act of 2002, Subtitle E • Animal Health Protection Act (PL 107-171), 21 U.S.C 114 • The Animal Industry Act of 1988 • The Plant Protection Act (7 U.S.C. 7701-7772, Section 431) and the ensuing regulations in Titles 7 and 9 of the Code of Federal Regulations	• Provides Secretary of USDA broad authority and discretion o To prevent, detect, control, and eradicate diseases of pests and animals o To promulgate regulations and take measures to prevent introduction and interstate dissemination of communicable diseases of livestock within the United States.[10,11] • Legal bases for APHIS' monitoring and surveillance programs[12]
	• 9CFR, Part 53, amended in 1985 (21 U.S.C., Section 151 et seq.)	• To respond to certain FADs and other communicable diseases of livestock or poultry and pay claims growing out of destruction of animal[13]
	• 9CFR subchapter B	• To establish cooperative programs to control and eradicate communicable diseases of livestock
	• 7 USC § 8303, 8304, 8305, 8306, 8307(b), 8308, 8310, 8312, 8314	• Statutory authority for emergency response — authorizes the Secretary to hold, seize, quarantine, treat, destroy, dispose of, or take other remedial action; declare an extraordinary emergency because of the presence of a pest or disease of livestock that threatens the livestock of the United States[14]
	• Virus-Serum Toxin Act of 1913	• APHIS-CVB to conduct inspection and compliance activities
	• The Foreign Service Act (1980) and Executive Order 12363 (1982)	• APHIS-IS activities

USDA, Agricultural Research Service (ARS)	• 5 U.S.C. 301 and Reorganization Plan No. 2 of 1953	• Established ARS on Nov 2, 1953
	• Title VII Section 1402 part 8 of the Farm Bill	• Authorizes the ARS food safety research programs to "maintain an adequate, nutritious, and safe supply of food to meet human nutritional needs and requirements."[15]
	• The Department of Agriculture Organic Act of 1862 (7 U.S.C 1 2201, 2204) • Research and Marketing Act of 1946, amended (7 U.S.C. 427k 1621) • Food and Agriculture Act of 1977, as amended (7 U.S.C. 1281 note) • Food Security Act of 1985 (7 U.S.C. 3101 note) • Food, Agriculture, Conservation, and Trade Act of 1990 (7 U.S.C. 1421 note) • Federal Agriculture Improvement and Reform Act of 1996 (FAIR Act) • The Agricultural Research, Extension, and Education Reform Act of 1998 (P.L. 105-185)	• Authorize research currently performed by ARS [16]
USDA, Cooperative State Research, Education and Extension Service (CSREES)	• National Agricultural Research, Extension, and Teaching Policy Act of 1977, Section 1433, Public Law 95-113, as amended; Section 1429, Public Law 97-98; Public Law 99-198; Public Law 101-624; Public Law 104-127; Public Law 105-185[17]	
USDA, Food Safety and Inspection Service (FSIS)	• Federal Meat Inspection Act, the Poultry Products Inspection Act • The Egg Products Inspection Act	• Authorizes FSIS to regulates red meat, poultry, and certain egg products[18]

continued

TABLE C-1 Continued

Agency	Legal and Regulatory Foundation	Authorities
DHHS-FDA	• Title III of the Bioterrorism Act	• Provide the HHS Secretary with new authorities to protect the nation's food supply against the threat of intentional contamination and other food-related emergencies. FDA expects up to 420,000 facilities to register under this requirement.[19]
DHHS-CDC	• Titles 8 and 42 of the U.S. Code and relevant supporting regulations, such as Interstate Quarantine (42 CFR 70) and Foreign Quarantine (42CFR71).	• Authorizes CDC- National Center for Infectious Diseases (NCID), Division of Global Migration and Quarantine to make and enforce regulations necessary to prevent the introduction, transmission or spread of communicable diseases from foreign countries into the United States.[20]
	• The Foreign Quarantine regulation (42 CFR Part 71.54, Etiologic Agents, Hosts, and Vectors)	• Governs the importation of hazardous materials (etiologic agents, vectors and materials containing etiologic agents • Importation into the United States must be accompanied by a U.S. Public Health Service importation permit.[21] • CDC regulations govern the importation of dogs, cats, turtles, monkeys, other animals, and animal products capable of causing human disease. • Under these regulatory authorities, CDC has established embargo animals and monkeypox virus, and of birds from specified Southeast Asian countries.[22]

DHS-FEMA	• 42 USC 5122(1), 42 USC 5192a &b, 5197b 44 CFR 206.38(d)	• Defines emergencies and provides legal authorities for FEMA activities in animal disease emergencies and other catastrophic events.
DOD	• 42 USC 5170c	• Authority to request the President to direct the Secretary of Defense to utilize DoD resources to perform emergency work essential for the preservation of life and property
DHS	• The Homeland Security Act of 2002	• Establishes DHS and its directorates

TABLE C-2 DHS Statutory Authorities and Responsibilities

Agency/Directorate	Statutory Authority[25]	Responsibilities
Bureau of Customs and Border Protection	Subtitle A, Title IV, of the Homeland Security Act of 2002 (6U.S.C 201 et seq.) establishment of Border and Transportation Security (CBP resides w/in this directorate)	• Responsible for controlling all of America's land, sea, and air borders; protects U.S. economic security by regulating and facilitating the lawful movement of goods and persons across U.S. borders. o Agricultural Quarantine Program (border inspection) (former USDA-APHIS)[26]
Emergency Preparedness and Response Directorate (FEMA)	Section 502 of the Homeland Security Act of 2002 (6 U.S.C. 312) Manages and coordinates the Federal response to major domestic disasters and emergencies of all types in accordance with the Robert T. Stafford Disaster Relief and Emergency Assistance Act.	• Established in 1979 as independent agency; transfer to DHS in 2003 • Major Divisions: Response, Recovery, Preparedness and Mitigations; U.S. Fire Administration; 10 Operations Regions • FEMA leads the federal government's role in preparing for, mitigating the effects of, responding to, and recovering from all domestic disasters whether natural or man-made, including acts of terror. • National Incident Management System (NIMS) Development and implementation of a single, coordinated, real-time national capability involving federal, state, and local governments, tribal nations, and citizens participation to deal with emergencies, disasters, and other incidents within the United States to support the President's National Strategy for Homeland Security • Incident Management Team (IMT) Act as the core, field-level response teams for major disasters, emergencies, and acts of terrorism. The IMT is the first federal response asset to be deployed in response to a major disaster, emergency, or act of terrorism, including the use of weapons of mass destruction. • Biodefense
Science and Technology Directorate	Title III of the Homeland Security Act of 2002 (6 U.S.C. 181 et seq)	• Provides leadership for directing, funding, and conducting research, development, test and evaluation and procurement of technology, and systems to prevent the importation of chemical, biological,

		radiological nuclear and related weapons and materials and to protect against and respond to terrorist threats. • Plum Island Animal Disease Center—$12.9M increase will be used to begin addressing safety/compliance security issues. • Bio-Surveillance Program Initiative—An increase of $65.M in FY05 is provided for a new initiative on bio-surveillance.
Information Analysis and Infrastructure Protection Directorate	Title II of the Homeland Security Act of 2002 (6 U.S.C. 121 et seq)	• Serves as focal point of intelligence analysis and infrastructure protection within the DHS. • Major divisions: Information Analysis Office, Infrastructure Protection Office; Homeland Security Operations Center (HSOC) • Assessments and Evaluations —Infrastructure Vulnerability and Risk Assessment • Funding to support: o development and maintenance of a complete and accurate mapping of the nation's critical infrastructure and key assets; o transportation and port vulnerability assessment; o specialized vulnerability assessment teams to conduct field assessments and catalog specific vulnerabilities of critical assets; o comprehensive national risk analysis activities, including modeling, data mining, and partnerships with scientific and academic communities to develop and refine DHS analytical tools and processes. • HSOC—a distributed system of centers, networks, processing and personnel that will provide the nation's single point of integration for homeland security information. The focus is federal, state, local, and private sector systems, and the HSOC is the principal mechanism for the execution of all DHS programs. • Biosurveillance Initiative—enable IAIP to integrate, in real time, biosurveillance data harvested through CDC, FDA, USDA, and DHS S&T. Combining this data with relevant threat information will enhance DHS' situational awareness of the health of U.S. population, animals, plants, food supply, and environment.

Table C-3 USDA Agencies and Programmatic Functions Addressing Animal Disease

Functions:
1) Deterrence & Prevention
2) Monitoring & Surveillance
3) Detection & Diagnosis
4) Education & Research
5) Communication

Agency	Description of Programmatic Activities		*1*	2	*3*	4	5
APHIS	Veterinary Service (VS)	• NCAHP (National Center for Animal Health Programs) initiates, leads, national certification & eradication programs to promote & improve U.S. animal health by preventing, minimizing or eradicating animal diseases of economic importance. • CVB (Center for Veterinary Biologics) assures safe, potent & effective veterinary biologics are available for diagnosis, prevention & treatment of animal diseases, involves in applied research or development activities related to veterinary biologics issues. • NCIE (National Center for Import and Export) facilitates international trade, monitors health of animals presented at border, regulates import & export of animals, animal products & biologics, diagnoses foreign & domestic animal diseases. • NVSL (National Animal Health Laboratory Network) provides diagnostic services (animal disease testing for VS). NVS lab in Ames, IA & FAD diagnostic lab on Plum Island. o NAHLN • CEAH (Centers for Epidemiology & Animal) develops technology applications, maintains key databases; conducts epidemiological, economic, & spatial analysis, and trade risk assessments; administers the NAHMS & other surveillance efforts. o NAHMS collects, analyses & disseminates data on animal health, management, & productivity across the United States. o CAHM (Center for Animal Health Monitoring) delivers timely information & knowledge about interactions among animal health, welfare, production, product wholesomeness, & the environments. o NAHRS (reporting system) is a joint effort of USAHA, AAVLD & APHIS • EP (Emergency Program) monitors foreign animal health and maintains surveillance at rapid detection and diagnosis of outbreaks of exotic diseases in the U.S. EP program.	X	X	X	X	X

	staff also enlists the help of more than 40,000 federally accredited veterinarians from the private sector who assist with disease exclusion and control. • CADIA (Center for Animal Disease Information & Analysis) collects, manages, & disseminate information critical to VS programs & other APHIS initiatives. • CEI (Center for Emerging Issues) works on emerging animal health issues keeps agency, industry, & the public informed. • EP (Emergency Program) provides training on Emergency Management Response System and outbreak symposiums.					
Wildlife Service (WS)	• Protects livestock from wildlife-borne diseases such as CWD. • Works with state counterparts to monitor wildlife diseases like rabies. • Has cooperative agreements with state wildlife agencies — Funds for surveillance and management distributed according to formula developed with International Association of Fish and Wildlife Agencies (IAFWA).[27] • National Wildlife Research Center functions as research arm. • Tests fish biologics at APHIS' NVS labs in Ames, IA, issues import permits for veterinary biologics, inspect veterinary biologics production facilities, methods & records. • Works with U.S. Dept. of the Interior's U.S. Fish and Wildlife Service (FWS), the U.S. Department of Commerce's National Marine Fisheries Service, and members of the Joint Subcommittee on Aquaculture (JSA) to disseminate information and outline agency roles to best meet the aquaculture industry's animal health needs. • Develops aquatic animal health monitoring and surveillance program. • Conducts research projects to study interaction of deer and cattle, coyotes as reservoirs of TB, and CWD in elk.	X	X	X	X	X
International Service (IS)	• Cooperates in major surveillance, eradication & control programs in foreign countries. • Monitors pest and diseases overseas.	X	X			
Agricultural Research Service (ARS)	• Conducts research to develop and transfer solutions to agricultural problems of high national priority; provides and disseminate information to ensure high-quality, safe food, and other agricultural products.[28] • 1,200 research projects organized into 22 national programs, including animal health, aquaculture and food safety.[29, 30, 31]				X	X

continued

Table C-3 Continued

Agency	Description of Programmatic Activities	1	2	3	4	5
	• Provides online educational materials for students, educators, & anyone interested in agricultural sciences. • 100 research locations, including overseas.[29] • National Animal Disease Center (NADC) is the major USDA center for research on livestock & poultry diseases.[32]					
Cooperative State Research, Education, and Extension Service (CSREES)	• Supports animal health and disease research at eligible schools and colleges of veterinary medicine, and state agricultural experiment stations.[33] • Established a unified network of public agricultural institutions to identify and respond to high-risk biological pathogens in the food and agricultural system. • National Research Competitive Grants Program (NRI) funds research on key problems relevant to agriculture, food, and the environment on a peer-reviewed, competitive basis.[34] • Partnering with the university system, supports expertise in plant and animal sciences along with extension and outreach capability that can be mobilized to provide an immediate response to critical issues. Program efforts will focus on early intervention strategies to prevent, manage, or eradicate new and emerging plant and animal diseases. Funding also will facilitate improved diagnostic tests for rapid response to emerging disease agents by expanding the knowledge base of microbial genomic for both animal and plant diseases.[103] • Supports research and extension base programs		X		X	X

Functions: 1) Deterrence & Prevention 2) Monitoring & Surveillance 3) Detection & Diagnosis 4) Education & Research 5) Communication

	o Provides resources to foster regional and national joint planning, encourage multi-state planning and program execution and minimize duplication of effort. o Funds, along with matching funds from the states, ensure responsiveness to emerging issues such as foot-and-mouth disease, *E.coli*, salmonella, listeria, sorghum ergot, potato late blight, and Russian wheat aphid.[103]					
Food Safety and Inspection Service (FSIS)	• Participates in Electronic Laboratory Exchange Network, or eLEXNET. This Internet-based system will be the mechanism by which the FERN laboratories report results from all bioterrorism or chemical terrorism related analyses. • FSIS participates in the CDC laboratory Response Network, provides training and microbiological methods to participants. • Reviews & approves foreign inspection systems & plants exporting meat, poultry, and processed egg products to the United States. • Operates three labs to perform scientific testing in support of inspection operations.	X	X	X	X	X

1.2.1.1 Border Strategy

On November 21, 2002, President Bush signed legislation creating the Department of Homeland Security (DHS) to unify federal forces and protect the nation from a new host of terrorist threats. Approximately 2,600 employees from APHIS' Agriculture Quarantine and Inspection (AQI) force became part of DHS' Border and Transportation Security's Bureau of Customs and Border Protection (CBP) on March 1, 2003.[37] This network of veterinary inspectors and animal health inspectors at all U.S. ports of entry is the first line of defense in identifying materials entering the United States that may be introducing foreign animal diseases. A summary of the programmatic elements and functions of the DHS' CBP is provided in Table C-4.

Although DHS is now responsible for protecting the nation's border, APHIS through risk assessment, pathway analysis, and rule making, continues to set agricultural policy, including specific quarantine, testing, and other conditions under which animals, animal products and veterinary biologics can be imported, which is then carried out by DHS.[38] At ports of entry, there are also USDA-APHIS-VS port veterinarians who inspect live animals at border ports and animals in quarantine until testing is completed. They are located at 43 VS office areas and report to the veterinarian in charge of the VS-Area Office.[39] With agricultural border inspectors now being a part of the DHS, VS has identified the need for developing new protocol for training and interacting with these inspectors and the need to work with DHS to implement improvements recommended in the Animal Health Safeguarding Review regarding exclusion activities in its strategic plan.[40]

The Foreign Quarantine regulation (42 CFR Part 71.54, Etiologic Agents, Hosts, and Vectors) governs the importation of hazardous materials (etiologic agents, vectors and materials containing etiologic agents. [51] CDC has established regulations that govern the importation of dogs, cats, turtles, monkeys, other animals, and animal products capable of causing human disease. Under these regulatory authorities, CDC has established embargo on monkeys and other animals that could carry the monkeypox virus and on birds from specified Southeast Asian countries.[52] At present, the CDC-National Center for Infectious Diseases (NCID), Division of Global Migration and Quarantine has quarantine stations in Atlanta, New York, Miami, Chicago, Los Angeles, San Francisco, Seattle, and Honolulu. The quarantine operations involve coordination of numerous agencies, including:[53]

* Epidemic Intelligence Service (EIS) and other parts of CDC
* State and local health departments
* Customs and Border Protection (DHS)

TABLE C-4 DHS, Border and Transportation Security, Bureau of Custom and Border Protection, and Components Addressing Animal Diseases[41–50]

Agencies	Function
Border and Transportation Security (BTS)	• The largest of the 5 DHS directorates. • Includes former U.S. Customs Service, border security function/enforcement division of INS, APHIS, Federal Law Enforcement Training Center, and the Transportation Security Administration. • Responsible for securing the nation's air, land, and sea borders. • Responsible for securing the nation's transportation systems and enforcing the nation's immigration laws.
Bureau of Custom and Border Protection (CBP)—(Commissioner Bonner)	• March 1, 2003, approximately 42,000 employees were transferred from U.S. Customs Service, INS, and APHIS to the new CBP, a new agency under the BTS directorate within the DHS. • Approximately 2,700 former USDA employees from the AQI program and APHIS were transferred into DHS. • Former APHIS-PPQ personnel at POE who were directly involved in terminal/plane inspections (100% time) were transferred to DHS; those with 60-70% time not doing inspection at terminals/planes were not transferred. • The agricultural import and entry inspection functions that were transferred include: reviewing passenger declarations and cargo manifests to target high-risk agricultural passengers or cargo shipments. • The new CBP also carries out the traditional missions of the predecessor agencies making up CBP (seizing illegal drugs & other contraband at the U.S. border; apprehending people who attempt to enter the United States illegally; detecting counterfeit entry documents; determining the admissibility of people and goods. protecting our agricultural interests from harmful pests or diseases; regulating and facilitating international trade; collecting duties and fees; enforcing all U.S. laws at our borders.

continued

TABLE C-4 Continued

Agencies	Function
Office Field Operations (OFO)	• Oversees over 25,000 employees at 20 field operation offices (OFOs), 317 ports of entry (POEs), and 14 pre-clearance stations in Canada and the Caribbean . • Responsible for enforcing customs, immigration, and agriculture laws and regulations at U.S. borders. • Manages core customs and border protection programs (i.e., border security and facilitation, interdiction and security, passenger operations, targeting analysis and canine enforcement; trade compliance and facilitation, trade risk management, enforcement, and seizures and penalties, as well as examines trade operations to focus on antiterrorism. • Administers Agricultural Inspection Policy and Programs (Agricultural Quarantine Inspection [AQI], at all POEs in order to protect the health of U.S. plant and animal resources) • Administers Immigrations policy programs. • Annual operating budget of $1.1 billion. • Each OFO is run by a Director of Field Operations (DFO).
OFO—Associate Commissioner of Agricultural Inspection Policy & Programs	• Policy advisor to the Office of the Commissioner on all agricultural issues.
CBP Port Director	• On March 1, 2003, CBP designated one Port Director at each POE in charge of all federal inspection services, establishing a single, unified chain of command.
CBP Ag. Specialist	• Enforces USDA regulations and seizes any articles in violation of those regulations. Conducts prearrival risk analysis. • Examines cargo for quarantine disease and pests. • Collects, prepares, and submits pest and disease samples to USDA.

	• Seizures, safeguarding, destruction, or reexportation of inadmissible cargo. • Negotiation of compliance agreements with importers of regulated commodities. • Stationed only at POEs with large volumes of cargo and only to support the CBP officers. • As of 10/4/2003, there are 1,471 full time permanent agricultural inspectors on board. • New CBP officers will be trained at the Federal Law Enforcement Training Center (FLETC) in Glynco, GA, and agricultural specialists will continue to learn their trade at PPQ Professional Development Center in Frederick, MD. • Farm groups and some members of congress have questioned whether CBP officers will receive enough agricultural training.
CBP & FDA	• In Oct 2003, CBP and FDA entered into an agreement to further protect the U.S. food supply. • At POEs, CBP inspectors now carry out special inspection & sampling of foreign food imports & make referrals back to FDA for further testing & analysis. • CBP and FDA work side by side in targeting efforts, making joint decisions about any food shipments that could pose a potential threat to the United States.
National Targeting Center (NTC)	• Part of CBP's OFO, the NTC provides tactical targeting and analytical research support for antiterrorism efforts to DHS and its operations center. NTC has representatives from all CBP disciplines.
CBP Laboratories and Scientific Sciences Division (LLS)	• On Dec 8, 2003, LLS moved its radiation portal monitor to the NTC.

* U.S. Department of Agriculture
* U.S. Fish and Wild Life Service
* The aircraft and maritime industry

The APHIS National Center for Import and Export (NCIE) also works to facilitate international trade, monitors health of animals presented at border, regulates import and export of animals, animal products and biologics, diagnoses foreign and domestic animal diseases. This APHIS center works in partnership with DOI's Fish and Wildlife Service, APHIS Plant and Protection Quarantine and DHS's CBP[54]

1.2.1.2 Offshore Strategy

Offshore activities are designed to mitigate pest and pathogen threats to the United States at points of origin. APHIS' International Service (IS), through international contacts, gathers and exchanges information on plant and animal health. APHIS-IS cooperates in major surveillance, eradication, and control programs in foreign countries, focusing on nations where economically significant pests or diseases are found. It has implemented the Offshore Pest Information System to monitor and document changes in distribution and outbreak status of specific, designated high risk exotic pest plants and animal diseases, including pathways, in their countries of origin. APHIS-IS currently has 64 foreign service officials stationed in 27 countries on six continents.[55] These APHIS personnel are engaged in surveillance and barrier programs, import and export trade facilitation, and commodity preclearance programs.[56]

USDA-ARS also operates six overseas locations for research on biological control of pests and pathogens. The research contributes to accurate identification of foreign pests and pathogen species; knowledge of basic biology; habitat characterization; assessment of ecological requirements; knowledge of limiting environmental conditions and patterns of occurrence; climate matching; and identification of potential control agents for foreign species. These programs target primarily unintentional threats. [57]

USDA-APHIS-VS' Center for Epidemiology and Animal Health (CEAH) is a collaborating center of the Office International des Epizooties (OIE) for Animal Disease Surveillance and Risk Analysis. The OIE serves as the world animal health organization. The VS' Center for Veterinary Biologics (CVB) and National Veterinary Services Laboratories (NVSL) participate as collaborating centers for the Diagnosis of Animal Diseases and Vaccine Evaluation in the Americas through their involvement in the Institute for International Cooperation in Animal Biologics. The NVSL also serves as an OIE reference laboratory for numerous diseases and provides training, consultation, and assistance to both domestic and international laboratories. As collaborating centers and reference laboratories, VS provides training, consultation, and other services to OIE members.[58,59]

1.2.1.3 Early Detection and Intelligence

In recent years, a few USDA staff have been detailed to intelligence and law-enforcement organizations. However, from the review of the these activities, the National Academy of Science has indicated in a 2003 report that it is unclear what information or approaches have been gleaned from these details, or whether findings have been incorporated or used by USDA.[60] Recent inquiries also showed that APHIS coordinates with DoD-Armed Forces Medical Intelligence Center (AFMIC) on intelligence information through detailing its intelligence analysts.[61] AFMIC is a field production activity of the Defense Intelligence Agency and the sole DoD producer of medical intelligence. The Center provides all source intelligence on worldwide infectious disease and environmental health risks. AFMIC maintains extensive databases; monitors foreign research, development, production, and transnational flow of medical materiel for military interest; and provides intelligence liaison services. APHIS intelligence analysts detailed at AFMIC review AFMIC data and coordinate monthly meetings on domestic and international threats for APHIS.[62]

The DHS Information Analysis and Infrastructure Protection (IAIP) Directorate gathers and assesses intelligence and information about threats and vulnerabilities from other agencies and takes preventive and protective action. The Department of Commerce's Critical Infrastructure Assurance Office (CIAO) and the FBI's National Infrastructure Protection Center were folded into this directorate. Agriculture and food are 2 of 14 critical infrastructure and key assets identified in the President's National Strategy for Homeland Security and thus fall into the domain of the IAIP Directorate.

Recently issued, the Homeland Security Presidential Directive/HSPD-9 establishes policy to defend the agriculture and food system against terrorist attacks, major disasters, and other emergencies.[63] The directive makes the DHS responsible for coordinating federal programs aimed at protecting U.S. agriculture and food from diseases, pests, and toxins. In coordination with the Secretaries of USDA, HHS, and the Administrator of EPA, the Attorney General, the Secretary of DHS, and the Director of CIA are to develop and enhance intelligence operations and analysis capabilities focusing on the agriculture, food, and water sectors (section 9, HSPD-9). The heads of Interior, Agriculture, Health and Human Services, the Administrator of EPA, and other agency heads are responsible for expanding the current monitoring and surveillance programs (section 8, HSPD-9) to develop:

- Robust, comprehensive, and fully coordinated surveillance and monitoring systems, including international information, for animal disease, plant disease, wildlife disease, food, public health, and water quality that provides early detection and awareness of disease, pest, or poisonous agents.
- Tracking system for specific animal, plants, commodities of food.
- Nationwide laboratory networks for food, veterinary, plant health, and water quality that integrate existing federal and state laboratory resources.

Additionally, the Secretary of DHS is directed to provide a report on options of creating a new biological threat awareness capacity that is based on this upgraded surveillance system to enhance detection and characterization of an attack. This report is not yet available at the preparation of this paper.

1.2.2 Monitoring and Surveillance

APHIS defines monitoring as the routine collection of information for a disease condition, characteristic, or state in an animal population; and surveillance as the analyses of the collected data.[64] A surveillance system that provides adequate early information about diseases and other animal health situations is crucial for rapid response. APHIS considers surveillance as the foundation for its Veterinary Services (VS) program activities, which include: domestic disease control and eradication programs, emergency preparedness, response, and trade. The APHIS Animal Health Monitoring Surveillance program is mandated by the Animal Health Protection Act and is conducted through partnerships with states, industry, and other federal agencies.

1.2.2.1 Current Animal Health Surveillance Program

The APHIS-VS Centers for Epidemiology and Animal Health (CEAH) in Fort Collins, Colorado, administers the National Animal Health Monitoring System (NAHMS) and other surveillance efforts. NAHMS collects, analyzes, and disseminates data on animal health, management, and productivity across the United States. APHIS officials collaborate with state and other federal agencies to conduct animal health surveillance activities, including: pre- and post-entry testing of imported animals, sample collection at slaughter, and routine testing of animals for export and interstate movement. APHIS also conducts surveillance for domestic animal

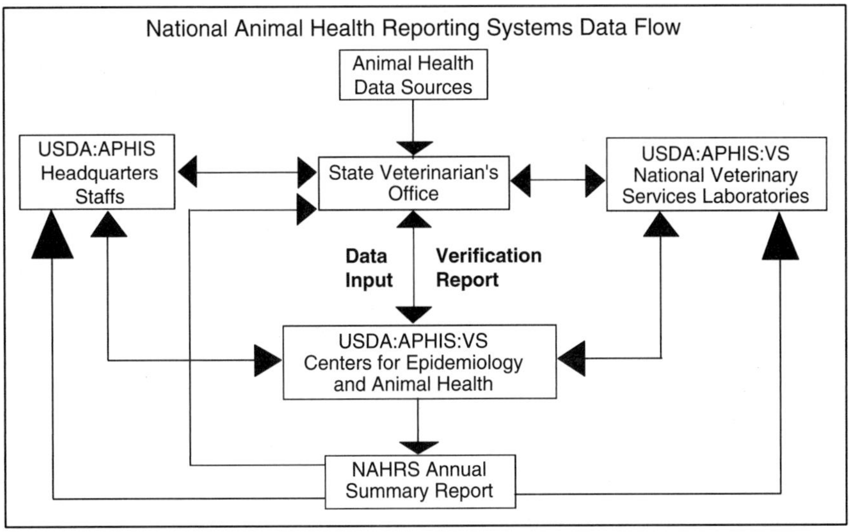

FIGURE C-2[68]

disease eradication programs, such as for brucellosis, tuberculosis, chronic wasting disease, and others.[65,66]

The National Animal Health Reporting System (NAHRS) is a joint effort of the U.S. Animal Health Association (USAHA), the American Association of Veterinary Laboratory Diagnosticians (AAVLD), and the USDA-APHIS. It is a reporting system designed to collect data on the presence of confirmed Office International des Epizooties (OIE) List A and B clinical diseases in commercial livestock, poultry, and aquaculture species in the United States. Using well defined reporting criteria, the chief animal health official of the state reports OIE List A and B clinical diseases for his or her state. USDA APHIS VS utilizes the data to complete monthly/annual animal disease status reports to OIE and to support trade negotiations. The report utilize multiple sources in reporting to OIE (the United States has been reporting for 25 years).[67] Figure C-2 describes the NAHRS data flow.

APHIS also conducts surveillance for early identification of foreign animal diseases (FAD). Surveillance is based on clinical symptoms (syndromic). Throughout the United States, APHIS has foreign animal disease diagnosticians (FADD) who are available within a 4-hour drive of any continental U.S. location. This is a network of 3-4,000 accredited veterinarians nationwide and about 400 state veterinarians.[69] The FADDs are especially trained to recognize, test, and diagnose FADs at Plum Island. Samples are collected and confirmatory tests are carried out. Tissue samples are sent to the DHS Plum Island Animal Disease Center or to the NVSL in Ames, Iowa, to rule out the presence of a FAD. The total number of FAD investigations in the United States has increased over the last few years. In FY 2002, VS FADDs conducted 837 investigations, an increase from 801 in FY 2001.[70] The scenario of a FAD involving live animals entering the United States is not likely, given the safeguards at ports of entry. More likely scenarios are passengers bringing in the virus inadvertently or intentionally.[71]

1.2.2.2 Animal Health Surveillance Enhancement at USDA

The demands of detecting foreign and emerging animal diseases, monitoring disease trends and threats in the United States and abroad, and providing adequate animal health information to various audiences require a flexible and dynamic surveillance system. Toward this goal, in October 2003, APHIS created the national surveillance unit (NSU) as the operational unit for the development of the National Surveillance System (NSS). It is a unit within the Center for National Surveillance at the VS-CEAH. Also created are several new positions to improve coordination of surveillance activities, including a National Surveillance Coordinator and a FSIS liaison. The main task of the appointed NSS Coordinator is to enhance and integrate national animal health surveillance to implement the enhancements recommended in the National Association of State Departments of Agriculture Animal Health Safeguarding Review and to facilitate effective interaction between APHIS and other federal agencies and stakeholders with surveillance activities. [72,73] Additional positions will be created at the NVSL and at the CEAH. In addition, several working groups, including a field implementation team, are being developed or are already in place.[74]

1.2.2.3 National Animal Identification and Tracking System

In collaboration with industry and stakeholders, VS is developing a National Animal Identification System to meet current and future animal health needs of American agriculture. The National Identification Development Team, a group of over 70 individuals representing over 30 organizations has developed the U.S. Animal Identification Plan (USAIP, see box). During the next 5 years, VS will also implement a new electronic permit system. This system will draw data from numerous sources, such as the ePermits system and Import/Export databases, by identifying key electronic linkages between these sources. This system should improve customer service and allow better tracking of products imported into the United States.[75,76]

The United States Animal Identification Plan (January 2004)[77] has evolved from the initial report, the National Identification Work Plan, that was presented at the 2002 USAHA meeting, with the recommendation that the USDA APHIS establish a joint state, federal, and industry group to further advance the work plan. Throughout 2003, approximately 100 animal and livestock industry professionals representing more than 70 associations, organizations, and government agencies, have formed the National Identification Development Team (NIDT) and worked to advance the work plan. This national identification and tracking plan will enhance disease preparedness by rapidly identifying animals exposed to disease, thus allowing quick detection, containment, and elimination of disease threats. When operational, the plan will be capable of tracing an animal or group of animals back to the herd or premises that is the most logical source of a disease concern. The plan's long term is to establish a system that can complete the traces (back and forward) within 48 hours of discovery of a disease. The USAIP intends to define the standards and framework for implementing and maintaining a national animal identification system for the US. The immediate priority is to have these standards recognized in the Code of Federal Regulations. The cattle, sheep, and swine industries have already developed preliminary implementation plans. All other livestock, including goats, cervids, equine, aquaculture, poultry, llamas, and bison, are becoming engaged in the plan.

1.2.3 Detection and Diagnosis

Early detection and reporting systems are key elements in a quick response in the event of an animal disease outbreak.[78] Much of the nation's expertise and the laboratories designed to make critical diagnoses of agriculturally important pests and pathogens are in the universities and USDA-Agricultural Research Services (ARS).[79]

USDA-APHIS-VS' National Veterinary Services Laboratories (NVSL) are the only federal laboratories dedicated to the testing of diagnostic specimens for domestic and foreign animal diseases. NVSL operates facilities in Ames, Iowa, and has diagnostic capabilities at Plum Island, New York (high security biocontainment FAD Diagnostic Lab, FADDL). On June 1, 2003, the property and facility of Plum Island Animal Disease Center, jointly run by USDA's APHIS and ARS, were transferred to DHS. NVSL analyzes blood, tissues, and environmental samples to promote disease tracking and identification.[80] VS continues to work with manufacturers, the ARS and other research agencies, and animal industry groups to facilitate the licensing of diagnostic tests used in the detection of foreign and emerging animal diseases.[81]

Although the authority for a federal response to animal diseases in the United States resides with APHIS, as delegated by the Secretary of Agriculture, the initial detection of new pathogen on a farm or ranch, preliminary diagnosis, and development of a program for its control relies heavily on collaborations with other groups, agencies, and individuals. Groups working together to ensure early detection and response to animal disease outbreaks are: USDA- Cooperative State Research, Education, and Extension Service (CSREES), Regional Emergency Animal Disease Eradication Organization (READEO), USDA-VS-NVSL (Ames, Iowa), DHS's Plum Island (FADDL), academics, professional societies, industry groups, other USDA agencies, other federal agencies, state departments of agriculture, state officials, and international organizations.[82,83]

In the past, VS has been able to rely on its close programmatic interface and share regulatory responsibilities with the state veterinarians to create consistency and standards in approaches taken to address animal health concerns. VS, however, has a different relationship with state fish and game agencies and with other federal agencies such as the DHS' Federal Emergency Management Agency (FEMA) and organizations such as the American Association of Veterinary Laboratory Diagnosticians (AAVLD). Additional models of cooperation will need to be developed and refined with these organizations to ensure consistency and define VS's leadership role.[84]

1.2.3.1 Laboratory Networks

1.2.3.1.1 National Animal Health Laboratory Network (NAHLN) In June 2002, the Public Health Security and Bioterrorism Preparedness and Response Act of 2002 was signed into law. Section 335 authorizes the Secretary of Agriculture to develop an agricultural early warning surveillance system enhancing the capacity and coordination among state veterinary diagnostic laboratories and federal and state facilities and public health agencies and provides authorization for Congress to appropriate funding to the NAHLN.[85] NAHLN addresses diagnostic needs for routine animal disease surveillance as well as diagnostic capacity for investigations and control and eradication programs.[86] The overall goal of the NAHLN is to contribute to the improvement of national disease surveillance capabilities. The concept was developed in discussion with NVSL that resulted in an MOU with AAVLD. The initial support was provided by CSREES and APHIS.[87]

The philosophy behind the design and implementation of NAHLN is that animal disease surveillance functions are most effectively accomplished as a shared responsibility among all animal health agencies. Under the NAHLN concept, state laboratories could provide significant surge capacity during a disease outbreak. The state labs could assist in defining herds for depopulation, delimiting the extent of the outbreak, and conducting follow-up surveillance to determine "free status."[88] The key goals of the NAHLN are to expand detection and response measures for pathogens that threaten animal agriculture and bolster laboratory capability for select agents with support for personnel, equipment, testing, and training. Among the elements planned for the NAHLN systems are to support the development and deployment of standard diagnostic approaches for identification of select agents, rapid diagnostic techniques, modern equipment, and experienced personnel trained in the detection of emergent, foreign, and bioteror agents; national training; proficiency testing, and quality assurance; and upgraded facilities meeting biocontainment requirements.[89] NAHLN is also to bolster data sharing among animal health agencies through the creation of a secure, two-way communications network and the creation of a national repository for animal health data; bolster cooperation and communication among animal health officials through maintenance of confidentiality of source data and providing alerts at appropriate response level.[90]

The network is currently in a pilot phase and modeled after the comprehensive response network in place for public health threats. The pilot NAHLN involves 12 state/university diagnostic laboratories to develop capacity and surveillance programs for eight high priority foreign animal diseases considered to be a bioterrorist threat (agent for foot-and-mouth disease, hog cholera, African swine fever, rinderpest, contagious bovine pleuropneumonia, lumpy skin disease, highly pathogenic influenza, exotic Newcastle disease). Other agents of interest for potential future inclusion include agents of zoonotic importance like West Nile encephalitis virus, Rift Valley fever, Nipah encephalitis virus, Hendra encephalitis virus, other encephalitides, and bovine spongiform encephalopathy.[91] Specifically, VS provided NAHLN laboratories with training in the standard nomenclature to be used in the pilot lab results reporting tool. VS anticipates that training in additional techniques will be offered and the number of NAHLN laboratories will increase significantly by FY 2009, attaining a broader pool of expertise to tap for surge testing capacity in an outbreak.[92]

1.2.3.1.2 The Laboratory Response Network (LRN) VS and the AAVLD are also partnering with the CDC to enlist state veterinary diagnostic laboratories into the CDC Laboratory Response Network (LRN). VS will serve as the gatekeeper for the veterinary diagnostic laboratory connection to this wider network. Figure C-3 outlines NAHLN structure with linkage to CDC-LRN. NVSL's Diagnostic Bacteriology Laboratory has received approval from the LRN to conduct diagnostic testing for *Clostridium botulinum*, *Francisella tularensis*, and *Yersinia pestis*. NVSL had previously been

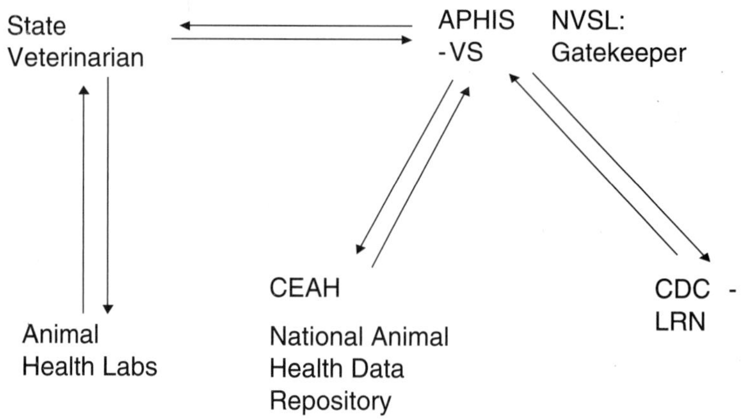

FIGURE C-3 NAHLN Structure.

approved for *Bacillus anthracis* and *Burkholderia spp*. The LRN laboratories function as confirmatory laboratories for other diagnostic laboratories and will process overflow samples in the event that a bioterrorist event were to occur.[93]

LRN was instituted in 1999 in preparation for the U.S. response to bioterrorism. The mission of LRN and its partners is to maintain an integrated national and international network of laboratories that is fully equipped to respond quickly to acts of chemical or biological terrorism, emerging infectious diseases, and public health threats and emergencies.[94] CDC runs the LRN program with direction and recommendations provided by the following agencies and organizations:[95]

- The Association of Public Health Laboratories
- The Federal Bureau of Investigation (Department of Justice)
- The American Association of Veterinary Laboratory Diagnosticians
- The American Society for Microbiology
- The EPA
- USDA
- DoD
- FDA
- DHS

LRN is a consortium of about 120 laboratories, which provide immediate and sustained laboratory testing and communication in the event of public health emergencies, particularly bioterrorism-related events. The network includes the following types of labs:[96]

• Federal—labs at CDC, the USDA, the FDA, and other facilities run by the federal agencies

• State and local public health—these are labs run by state and local departments of health

• Military—labs operated by the Department of Defense, including the U.S. Army Medical Research Institute for Infectious Diseases (USAMRIID) at Fort Detrick, Maryland

• Food testing—the LRN includes FDA labs and others that are responsible for ensuring the safety of the food supply

• Environmental—includes labs that are capable of testing water and other environmental samples

• Veterinary—some LRN labs, such as those run by USDA, are responsible for animal testing. Some diseases can be shared by humans and animals, and animals often provide the first sign of disease outbreak.

• International—the LRN has labs located in Canada, the United Kingdom, and Australia

The LRN labs are designated as either national, reference, or sentinel. The structure is as follows:[97]

• National labs include CDC and USARMRIID labs with unique resources to handle highly infectious agents and the ability to identify specific agent strains.[98]

• Reference labs, sometimes referred to as confirmatory reference, can perform tests to detect and confirm the presence of a threat agent. These labs ensure a timely local response, rather than having to rely on confirmation from labs at CDC, allowing quick local response.

• Sentinel labs represent the thousands of hospital-based labs that are in the front lines. Their responsibility is to refer a suspicious sample to the right reference lab.

For example, in the case of severe acute respiratory syndrome (SARS), the CDC labs identified the unique DNA sequence of the virus that causes the disease. LRN developed tests and materials needed to support these tests and gave LRN members access to the tests and materials.[99]

1.2.4 Research, Education and Training

Research and education programs that are needed for managing and recovering from a new pest and pathogen typically involve a land-grant university and ARS programs, state departments of agriculture, state officials, and APHIS.[100]

1.2.4.1 USDA-Cooperative State Research, Education, and Extension Service (CSREES)[101]

CSREES distributes money to geographically disperse state laboratories to fund facility and equipment upgrades; participates in a nationwide system of agricultural research and education program planning and coordination between state institutions and the USDA; assists in maintaining cooperation among the state institutions, between the state institutions and their federal research partners; administers grants and payments to state institutions to supplement State and local funding for agricultural research and higher educations; funds animal health and disease research by formula to support livestock and poultry disease research in 67 colleges of veterinary medicine and in eligible agricultural experiment stations.

Under the National Initiative competitive grants, CSREES supports research in plants and animals; natural resources and the environment; nutrition; food safety; health; markets, trade, and rural development; and processing for adding value or developing new products.

CSREES's objective is to support animal health and disease research at eligible schools and colleges of veterinary medicine and state agricultural experiment stations whose purpose is to improve the health and productivity of food animals and horses through effective prevention, control, or treatment of disease, reduction of losses from transportation and other hazards, and protect human health through control of animal diseases transmissible to people.[102]

Through cooperative efforts with APHIS, CSREES established a unified network of public agricultural institutions to identify and respond to high-risk biological pathogens in the food and agricultural system. This program develops and promotes curricula for higher education programs to support the protection of animals, plants, and public health; support interdisciplinary degree programs that combine training in food and agricultural sciences, medicine, veterinary medicine, epidemiology, microbiology, chemistry, engineering, and mathematics (statistical modeling).[103]

Under the National Research Competitive Grants Program (NRI), CSREES funds research on key problems relevant to agriculture, food, and the environment on a peer-reviewed, competitive basis. It was established in 1991 in response to recommendations outlined in *Investing in Research: A Proposal to Strengthen the Agricultural, Food and Environmental System*, a 1989 NRC report.[104] Partnering with the university system, CSREES programs support expertise in plant and animal sciences along with extension and outreach capability that can be mobilized to provide an immediate response to critical issues. Program efforts will focus on early intervention strategies to prevent, manage, or eradicate new and emerging plant and animal diseases. Funding also will facilitate improved diagnostic tests for rapid response to emerging disease agents by expanding the knowledge base of microbial genomics for both animal and plant diseases.[105]

CSREES provides resources necessary to foster regional and national joint planning, encourage multistate planning and program execution, and minimize duplication of effort. In order to assure responsiveness to emerging issues such as foot-and-mouth disease, *E. coli, Salmonella, Listeria,* sorghum ergot, potato late blight, etc. CSREES provides funds along with matching funds from the states. It has provided $43 million to states, university, and tribal lands to increase homeland security prevention, detection, and response efforts.[106]

1.2.4.2 USDA-Agricultural Research Service (ARS)

Over 1,200 USDA-ARS research projects are organized into 22 national programs and three major areas of focus: Animal Production, Product Value and Safety; Natural Resources and Sustainable Agricultural Systems; Crop Production, Product Value, and Safety. Under the Animal focus is the Animal Health Program.[107] The mission of ARS-National Animal Health Program is to conduct basic and applied research on selected diseases of economic importance to the U.S. livestock and poultry industries.[108] The research is intended to provide scientific information for the control or elimination of animal diseases, optimize animal production systems, and help to ease problems relating to sanitary regulations and food safety disputes.

The ARS-National Animal Health Program, Pathogen Detection and Diagnostics Component is intended to produce a new generation of diagnostic tools that will facilitate detection and identification of known pathogens and diseases, new variants of infectious and noninfectious agents, and emerging organisms and diseases.[109] USDA-ARS, National Animal Disease Center (NADC) in Ames, Iowa, is the major federal center for domestic and emerging animal disease research. The NADC opened in 1961 and consists of more than 80 buildings on 318 acres, with an annual budget of $23M, and is staffed with 300 people, including 56 scientists.[110] ARS's other research locations include:[111]

- Animal Disease Research Unit, Pullman, Wash.
- Arthropod-Born Animal Diseases Research Laboratory, Laramie, Wyo.
- Avian Diseases and Oncology Laboratory, East Lansing, Mich.
- Beltsville Agricultural Research Center, Beltsville, Md.
- Plum Island Animal Disease Center, Orient Point, N.Y.
- Poisonous Plant Research Laboratory, North Logan, Utah

- Poultry Research Unite, Mississippi State, Miss.
- Roman L. Hruska U.S. Meat Animal Research Center, Clay Center, Neb.
- Southeast Poultry Research Laboratory, Athens, Ga.

Most ARS animal health programs are oriented toward understanding the role of individual agents in disease or animal specific metabolic problems. However, many current and future disease problems have a multifactorial etiology, exist in a subclinical or chronic state, and production losses are not always obvious. Research on these complex problems requires multidisciplinary, multivariate dynamic analysis of real life field situations, such as epidemiology. However, there is little epidemiological expertise within ARS research programs. Although epidemiology expertise in APHIS is a valuable resource, it is mostly devoted to disease monitoring and other APHIS programs, not hypothesis driven for research purposes. ARS has indicated in its action plan that its goal is to incorporate hypothesis driven epidemiological analysis into selected current and future ARS research programs and to establish cooperative agreements with the APHIS-CEAH to promote ARS conducting experimental epidemiology research.[112]

1.2.4.3 DHS-Science and Technology (S&T) Directorate[113,114]

This directorate coordinates DHS efforts in research and development, including preparing and responding to the full range of terrorist threats involving weapons of mass destruction. It conducts research on models, simulations, and tabletop exercises designed to:

- Explore epidemiological and economic consequences
- Analyze research and development requirements for foreign animal disease and food security scenarios
- Develop key technologies and tools to prevent, detect, respond, and recover from intentional and unintentional introduction of biological agents to the national agricultural and food systems

The S&T plans to conduct systems studies to explore the potential utility of technology such as BioWatch for agricultural scenarios. Its strategy is to overlay protection form agricultural terrorism on the existing research and regulatory programs at USDA and FDA. Two of the four high-consequence biological scenarios that constitute the research programs for S&T Biological and Chemical Countermeasures Portfolio address major concerns for agriculture and food, mainly, the deliberate introduction of foot-and-mouth disease into the United States and a classified food security event.

Plum Island Animal Disease Center (PIADC) June 1, 2003, the S&T Directorate assumed responsibility for the operation of PIADC as mandated by the Homeland Security Act of 2002 (PL 107-296 Section 310). PIADC conducts research to protect the health of livestock across America from high-consequence foreign animal diseases such as foot-and-mouth disease. PIADC continues its veterinary sciences and animal health research under the transition:

- Research on foreign animal diseases that are not present in the United States (current focus on FMD and African swine fever)
- Developing a joint DHS/USDA comprehensive national strategy for foreign animal disease with emphasis on FMD
- International collaboration and commercialization of vaccines and diagnostic tools
- Work on zoonotic agents
- Biocontainment facilities operate at a biosafety level of 3

DHS and USDA have joint management over PIADC. USDA-ARS and APHIS continues their research and development of diagnostic programs, respectively. USDA research staff remains with USDA while contributing to PIADC's expanded agroterrorism mission. DHS assumed administration and management responsibilities for the facility.

Office of Research and Development (ORD)-National Biodefense Analysis Countermeasure Center (NBACC) NBACC is dedicated to protect health and agriculture by advancing the scientific community's knowledge of bioterrorism threats and vulnerabilities. NBACC integrates facilities and technical expertise in biodefense through a hub and spoke structure. The NBACC hub is based on the National Biodefense Campus at Fort Detrick in Maryland. The PIADC is one of the spokes. Additional operations are at the DOE National Laboratories. The Center serves three main functions:

- Provides scientific assessment of emerging biological threats both to people and to agriculture
- In partnership with the FBI, establishes new national capability for performing forensic analyses on bioterror events
- Works with the PIADC to conduct research to help defend against agroterrorism

University-Based Homeland Security Centers of Excellence (HS-Centers)[115] In December 2003, the S&T Division released a broad agency announcement calling for proposals that will focus on research effort to combat agroterrorism. DHS intends to establish two HS-Centers by April 2004; one will focus on animal related agroterrorism and the other on postharvest food security.

1.2.4.4 DHHS-NIH

Within HHS, a nontrivial amount of food safety research is funded by the National Institutes of Health (NIH). It is unclear, however, whether this research is coordinated with, or even complements, the research conducted by USDA, the FDA, and the EPA.[116]

1.2.4.5 DoD-U.S. Army Medical Research, Institute of Infectious Diseases (USAMRIID)

USAMRIID conducts research to develop strategies, products, information, procedures, and training programs for medical defense against biological warfare threats and infectious diseases. USARMRIID, an organization of the U.S. Army Medical Research and Materiel Command (USAMRMC), is the lead medical research laboratory for the U.S. Biological Defense Research Program. The institute plays a key role in national defense and in infectious disease research as the largest biological containment laboratory in the DoD for the study of hazardous diseases. USAMRIID has over 10,000 square feet of Biosafety Level 4 (BSL-4) and 50,000 square fee of BSL-3 lab space. Its 450 civilian staffs include veterinarians. Veterinary medicine is one of the major divisions of the Institute. Its current studies include work on improving vaccines for anthrax, Venezuelan equine encephalitis, plague, and botulism, and on new vaccines for toxins such as staphylococcal enterotoxins and ricin. Research on medical countermeasures to viral hemorrhagic fevers and arboviral illnesses also is conducted. A significant effort is devoted to developing both laboratory and field diagnostic assays for agents considered to be biological warfare or endemic disease threats.[117]

It has been previously suggested that research related to agricultural bioterrorism be expanded in non-USDA agencies with relevant capabilities or responsibilities, such as the FDA, CDC, USAMRIID, and Defense Advanced Research Projects Agency (DARPA). DARPA does not presently have the legislative authority to support agricultural research.[118]

1.2.5 Emergency Response and Communication

The Homeland Security Presidential Directive #5 (HSPD-5) enables the Department of Homeland Security to assume responsibility for coordinating federal response operations under certain circumstances. In particular, the DHS secretary will coordinate the federal government's resources in response to or recovery from terrorist attacks, major disasters, or other emergencies, when any one of the following conditions applies:[119]

- The federal department or agency acting under its own authority has requested assistance.
- The resources of state and local authorities are overwhelmed and federal assistance has been formally requested by states/local authorities.
- More than one federal agency has become substantially involved in responding to the incident.
- The secretary has been directed to assume responsibility for managing the domestic incident by the president.

HSPD-5 describes the National Incident Management System (NIMS), which is to cover all incidents, natural or unnatural, for which the federal government deploys emergency response assets. Under this directive, the Secretary of DHS is responsible for leading the development and implementation of NIMS.[120]

As directed by HSPD-5, APHIS has structured its Emergency Management Response Systems (EMS) systems according to NIMS. APHIS' EMS is a joint federal-state-industry effort to improve the ability to deal successfully with animal health emergencies, ranging from natural disasters to introductions of foreign animal diseases. The EMS program identifies national infrastructure needs for anticipating, preventing, mitigating, responding to, and recovering from such emergencies. In FY 2003 APHIS established five incident command posts in three states in response to the exotic Newcastle disease (END) outbreak in California, Arizona, Nevada, and Texas.[121] Recent APHIS' efforts to build emergency management capacity have included:[122]

1. Develop and implement plans for a secure operation in Riverdale, Md.
2. Continually update two Regional Emergency Animal Disease Eradication Organization (READEO) units capable of addressing an animal health emergency and interacting with the APHIS operations center in Riverdale, Md.
3. APHIS, with the NAHEMS Steering Committee, developed strategy for funding the construction of new emergency management

biocontainment and laboratory facilities at Plum Island and in Ames, Iowa.

1.2.5.1 Existing Federal Emergency Response Plan

USDA-APHIS-VS*—National Animal Health Emergency Response Plan For An Outbreak of Foot-And-Mouth Disease or Other Highly Contagious Animal Diseases*[123] outlines the national organization and concepts of operation for responding to a widespread highly contagious disease of animals. Much of the information provided in this section came from this plan. Specifically, this plan provides a unified response to all aspects of an FMD outbreak and primarily addresses the coordination and resources that would be required in a multiple-state outbreak. While the plan was written to address a widespread outbreak of FMD, its operating principals would also apply to large-scale outbreaks of other diseases, including, but not limited to, highly pathogenic avian influenza, Newcastle disease, classical swine fever, and African swine fever. Response for an outbreak of highly contagious zoonotic disease (transmitted to people from animals) may be addressed under this plan or may be addressed under contingency plans of the national human health system but will be coordinated under the National Response Plan.[124]

Depending on the scope of the situation, USDA will ask the Department of Homeland Security's Federal Emergency Management Agency (FEMA) to coordinate logistical response functions with other federal agencies and state(s) where FMD exists. FEMA would use the framework of the National Response Plan (NRP) in accordance with operational requirements and priorities established by agricultural authorities. (A memorandum of agreement [MOA] will be established between USDA and FEMA to outline respective roles and responsibilities in the event of a major FMD outbreak). The operational structure of the Federal Response Plan (FRP) will be utilized, with or without a presidential emergency or major disaster declaration, to provide an established federal/state coordination mechanism. (Currently an agricultural emergency is not covered under a Stafford Act Declaration.)

When the FMD outbreak or threat is intentional, USDA's Office of the Inspector General (OIG) would be notified. As warranted by the situation, OIG will then notify and coordinate with the appropriate law enforcement agencies at the local, state, and federal levels. If there is a suspicion that the outbreak was caused by criminal activity, the OIG would work closely with the responding veterinary staff for the proper handling, packaging, and shipment of any samples to the appropriate research laboratory for testing and forensic analysis. OIG would conduct any subsequent criminal investigation. If an FMD outbreak is determined to be a criminal but not terrorist act, OIG will assume federal lead responsibility for a law enforcement response. If an FMD outbreak is the result of a terrorist act, OIG would notify the Federal Bureau of Investigation (FBI), Weapons of Mass Destruction Unit. OIG and the FBI would jointly conduct a criminal investigation. Well-coordinated interagency mechanisms must be established among the FBI, USDA, and DoD for collaborative forensic investigations.[125]

1.2.5.2 Federal Response to a FMD outbreak or similarly infectious diseases[126]

If an FAD passes the first line of defense, responses are conducted through six APHIS-VS national incident management teams (comprising of federal employees) and one incident management team in each VS area office (comprising both federal and state employees). Additional support for the national response would be from the FEMA National Management System, and would be coordinated out of DHS. There is a USDA liaison (at present, the APHIS-VS Associate Deputy Administrator for Emergency Management) to DHS Federal Emergency Management Agency.[127]

USDA is the lead agency for all agricultural emergencies at the federal level. Within USDA, APHIS is the lead agency for managing an animal disease outbreak such as FMD and within APHIS, the VS-EP (Emergency Program) monitors foreign animal health and maintains surveillance, detection, and diagnosis of outbreaks of exotic diseases in the United States. In order to effectively deal with animal disease emergencies, coordinated response among a number of local producers, veterinarians in private clinical practice, and state-level veterinarians and animal health organizations are required. The APHIS-VS-EP also enlists the help of more than 40,000 federally accredited veterinarians from the private sector who assist with disease exclusion and control.

The initial response to an animal disease outbreak involves the existing USDA expertise, resources, and authorities. The authority for establishing the national strategic response policy for containment and eradication of an FMD outbreak reside in the National Incident Coordinator (NIC) (usually the APHIS-VS- Associate Deputy Administrator for Emergency Management). The USDA-NIC coordinates activities through the APHIS Emergency Operations Center (AEOC) in Riverdale, Maryland. USDA-APHIS Senior Public Affairs Officer establishes a Joint Information Center (JIC), collocated with the AEOC, to serve as the principal source of information and coordinate with other federal agencies, industry communications officials, and state-level JICs.[128]

APHIS opened the AEOC in March 2003. The center serves as the national command and coordination center for APHIS emergency programs disaster management. Teams working in the AEOC (the National Response Management Team, NRMT) have enhanced ability to collect, analyze and disseminate information. The NRMT can direct necessary resources and communicate with appropriate stakeholders by coordinating with other federal, state, and international organizations, including the Department of Homeland Security. Communication capabilities include video-teleconferencing, advanced computer interfaces, geographical information system mapping, and a strong multimedia component.[129, 130, 131]

At the regional level, in each of the two APHIS-VS regions (Eastern and Western), the APHIS Regional Director manages regional VS resources and communications. APHIS-VS has a disease eradication team made up of APHIS employees. These team members, as an additional duty, train for a specific function of disease response and will deploy individually or as a group in support of the affected state or states based on their needs. The state veterinarian can request that the APHIS area veterinarian in charge (AVIC) deploy this resource.[132]

In each infected state, the assigned AVIC is the authorized representative of USDA and is the senior federal official who interacts with the state coordinating officer (SCO). The SCO is the designated lead state official and may be the state veterinarian or some other official from either the agricultural or emergency management community. The AVIC leads the overall federal component of the integrated response within the state and establishes operational requirements. The AVIC and the SCO coordinate all activities at the state level.

In the field, the assigned incident commander (IC) will manage all response operations at an infected site. The IC may be a state or federal animal health official as designated by the SCO/AVIC team.

The Stafford Act, DHS-FEMA[133] Under the Stafford Act, FEMA serves as the coordinating agency for disaster response and recovery activities. Without a Stafford Act declaration, USDA can request FEMA and other federal agencies to provide support in performing traditional emergency management functions using the framework of the Federal Response Plan. In the event the President declares an emergency or major disaster under the Stafford Act, FEMA assigns a primary federal official (PFO) and the PFO. The PFO, in coordination with the SCO and AVIC, oversees federal response support operations authorized under the presidential emergency or major disaster declaration, and the AVIC continues to represent APHIS for containment and eradication operations. At FEMA headquarters, the FEMA Assistant Director for Readiness, Response, and Recovery provides general direction and oversight for FEMA support of USDA. FEMA activates the interagency emergency support team (EST) located at the National Interagency Emergency Operations Center (NIEOC) and coordinates with the Homeland Security Operations Center in Washington, D.C. To facilitate coordination of animal health and emergency management response functions, a USDA liaison reports to the NIEOC at FEMA headquarters and a FEMA liaison reports to the AEOC.

In the event of an agroterror attack, DHS leads the team of first responders to contain and manage the threat while APHIS provides crucial scientific and diagnostic expertise. APHIS' expertise is used in managing a potential disease outbreaks as well as in assisting DHS in its investigative and intelligence-gathering efforts to find those responsible for the terrorist attack.[134] Figure C-4 outlines the infrastructure for a federal response to an FMD outbreak.

Other Federal Agencies Other possible federal agency resources available for response to an FMD outbreak in the United States include:

- DOT for logistical, quarantine, decontamination, and animal carcass disposal; prevention of vessels with suspected FMD cargoes from entering U.S. waters; provision ships, planes, helicopters, and communication systems.
- National Communications System (NCS).
- U.S. Army Corps of Engineers for FMD debris and disposal, decontamination procedures, technical assistance in environmental site assessments, ground water monitoring, soil sampling, contract preparation, and GIS mapping.
- USDA/U.S. Forest Service can ensure that carcasses are disposed of in accordance with state law or local ordinance; enforcement of federal or state quarantine regulations.
- American Red Cross can provide hotline for affected farmers to request assistance and for concerned citizens to get information on activities, current conditions, and referrals to other relief agencies; community education to prevent disease spread.
- General Services Administration (GSA) can provide a number of assists, including:
    - o provide contract support for risk assessment; public education; monitoring; surveillance; detection; testing/diagnosis services; epidemiology; biosecurity; appraisal; vaccination; depopulation/disposal; cleaning and disinfections (site/premises); decontamination of equipment; public information and rumor control.
    - o GIS mapping services; environmental monitoring and plume projections; feeding and lodging (mass care) of support response personnel; computer equipment, support, and operations in response to federal operations; transportation services for team personnel and

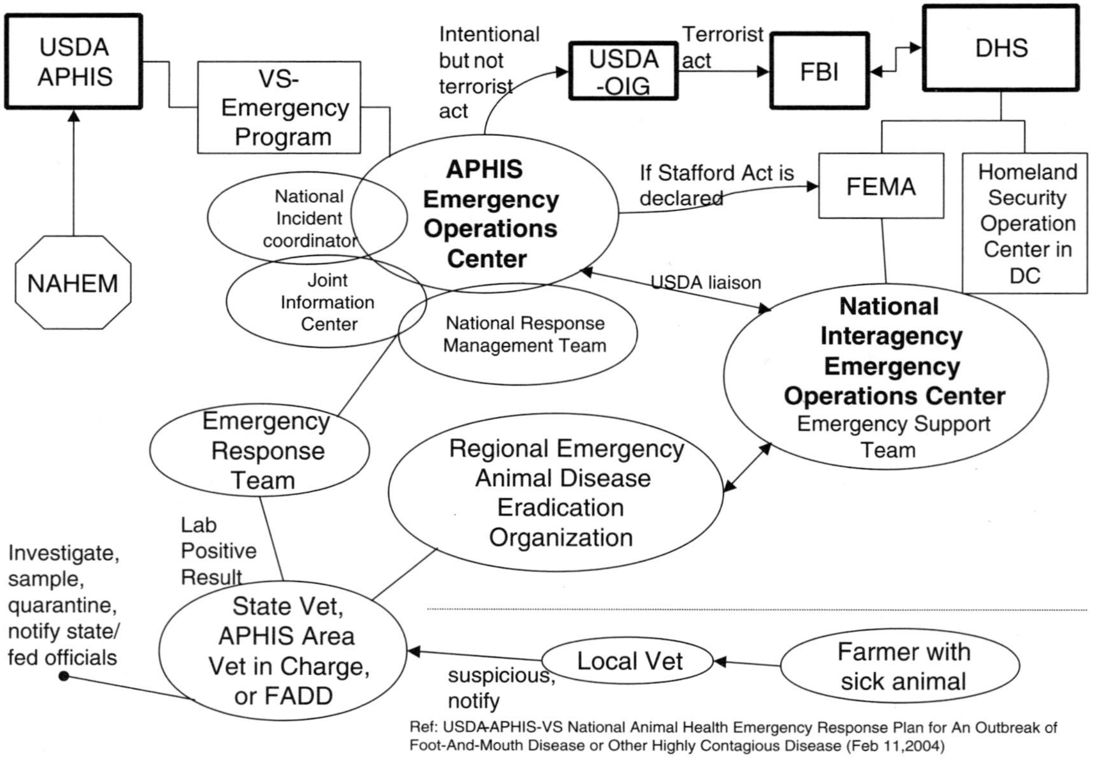

FIGURE C-4 Infrastructure for federal response to a FMD outbreak.

hazardous waste; public health and safety; mental health counseling; and temporary personnel and services.

o There exists an MOA between GSA and USDA-APHIS detailing the mechanism for requesting and receiving support in the event that the presence of animal/plant disease and/or pests constitutes an actual or potential emergency situation.

- DHHS:

o The FDA can provide food safety assistance, technical assistance, vaccination assistance, veterinary surveillance, carcass disposal assistance, direct animal care and assist in analyzing potential control issues; regulate domestic and imported animal feed products to prevent or curtail FMD; assist in environmental health, disease surveillance, and vector control regarding food and feed.

o The CDC can assist in disease surveillance, epidemiological and pest management, environmental risk assessment, worker safety issues and the provision of veterinary/medical personnel.

o The NIH can assist in the assessment of environmental impacts, disease surveillance, vector control issues and the provision of veterinary/medical personnel and has a grant program that can assist laboratories for addressing FMD.

- EPA can provide oversight in the decontamination effort by ensuring chemicals are contained and disposed to prevent environmental contamination; can review decontamination, chemical, and disposal plans; environmental air and water sampling.
- USDA, Food and Nutrition Service (FNS) educates the public about the safety of the food supply.
- DOE, through its atmospheric release advisory capability (ARAC), can model the spread of the FMD virus as well as the dispersion of smoke from burning debris.
- DOC-NOAA can provide weather forecast and observational data upon request; provide on-site meteorologists to support incident command operations; provide dispersion wind forecasts in coordination with other elements of NOAA; broadcast time-critical FMD information on NOAA Weather Radio.

1.2.5.3 Communication

Communication is an important part of an emergency management system. While communication flow through APHIS National Emergency Operation Center at Riverdale for all national coordination during a disease outbreak is well detailed, a written plan for coordination and communication of information that flows from the coordinating bodies at DHS does not appear to exist.[135]

1.3 BUDGET AND CAPACITY

Budget information and existing capacity at key USDA agencies with programmatic activities that address animal health are summarized in Table C-5. Budget information for the components of DHS with functions that may include functions such as early detection, prevention, and management of incidents involving animal diseases is summarized in Table C-6. APHIS FY 2005 budget for various programs addressing animal diseases reflects request for increase in capacity in the following areas:[136]

Program Areas	Requested Budget Increase FY 2005$
• Animal Health Monitoring and Surveillance increase: to support and enhance BSE surveillance	8.641M
• National Animal Identification program	33.197M
• State Cooperators	6.1M
• Biosurveillance program: to enhance data collection system, improve surveillance capabilities and establish connectivity with the integration and analysis function at DHS. Increase also allow increase in the number of FAD tests at the NVSL and approved state labs	5M
• Low pathogenic avian influenza program: to conduct surveillance and control program	11.783M
• FMD/FAD program: to reduce domestic threats through increase offshore threat assessment, including more officers overseas	4.229M
• Emergency Management Systems program	10.625M
• Veterinary biologics	1.861M
• Veterinary diagnostics program: to enhance the NAHLN and continues its diagnostic work at the FADDL on Plum Island to help protect U.S. herd against potential acts of bioterrorism	4.347M
• Import/Export program: to develop and implement an automated system to track animal and animal product movements	1.355M

1.3.1 Laboratory Capacity

Adequate space at the appropriate biosafety level is essential to conducting diagnostic activities. With DoD appropriations, VS began construction in December 2002 on a facility to relocate laboratories from leased space to the main APHIS site at Ames, Iowa. With other appropriate funds, construction began in the fall 2003 for the High Containment (BSL-3AG) Large Animal Housing Facility, associated with infrastructure, and miscellaneous support structures, which are targeted for completion in 2006. In addition, planning/design are well underway for the remainder of the National Cen-

TABLE C-5 Capacity and Level of Funding for USDA Agencies Addressing Animal Diseases

Agency	Overall Capacity and Funding	Diagnostic, Laboratory Network, Biocontainment Level	Surveillance and Monitoring	Deterrence and Prevention	Risk Analysis
APHIS	Overall APHIS: • Jan 2002 Defense Appropriation Act: $105M for pest and disease exclusion; $80M for upgrading USDA facilities for operational security • APHIS spent well over $100 million on disease diagnostic and epidemiology and pest detection infrastructure.[137] Veterinary Service (VS): • Approximately 1,600 permanent employees • Nationally distributed with field offices in each of the 50 states & major ports of entry. • Operating funds approx $210M (FY'03) VS-CVB operates on a $11M annual budget and has 101 employees, including 48 scientists. The CVB and NVSL shared some personnel.[138]	• The NVSL has a $15 M annual operating budget and 210 employees, of whom 60 are scientists.[139] • $20.6 M for national animal and plant diagnostic laboratory networks (2002 Homeland Security Supp.Fund)[140] • $14 M for increased security at NVSL in Ames, Iowa, and $23M for Plum Island Animal Disease Center (Jan 2002 Defense Appropriations Act) • Few facilities crucial for accurate diagnoses exist nationwide. Those in existence have limited resources to receive, analyze, and identify many potential agroterrorist agents. The NAHLN remains in pilot stage with laboratories in only 12 states. Further, these labs lack the capacity to test for more than 8 of the 37 FAD agents.[141] • $381M (Requested FY2005); of this o $178M is for NCAH (Ames, IA). o Only $30 M is requested for both plant and animal laboratory upgrades elsewhere.	• $138M (Est. FY'04, total APHIS plant & animal health monitoring)[144] • $134M (Actual FY'03, total APHIS plant & animal health monitoring)[143] • $219M (Est. FY'05, total APHIS plant & animal health monitoring; approx $36M is for BSE)[145]	• $351M (Actual FY'03, pest and disease exclusion)[146] • $285M (Est. FY'04, pest and disease exclusion) [147] • $315M (Est. FY'05, pest and disease exclusion) [148]	APHIS has a statistically based risk-assessment system to evaluate effectiveness of ongoing inspection operations.

continued

TABLE C-5 Continued

Agency	Overall Capacity and Funding	Diagnostic, Laboratory network, Biocontainment Level	Surveillance and Monitoring	Deterrence and Prevention	Risk analysis
		o The American Association of Veterinary Laboratory Diagnosticians estimated that at least an additional $85 M above the current funding is required to expand the network[142]			
APHIS-WS	• $64M (Estimate FY'03) • $66M (Budget FY'04)				
Agricultural Research Service (ARS)	• 100 research locations, a few locations overseas • 2,100 scientists and 6,000 other employees • Protection of livestock from diseases/ pathogens:[149] o $59M(Actual FY'03) o $65M (Est. FY'04) o $61M (Est. FY'05) • Defending against catastrophic threat, homeland security:[150] o $23M (Actual. FY'03) o $21M (Est. FY'04) o $49M (Est. FY'05)				
Food Safety and Inspection Service (FSIS)	• $15M for security upgrades and bioterrorism protection (Jan 2002 Defense Appropriation Act)[151] • $754M (Actual FY'03)[152] • $775M (Est. FY'04) [153] • $839M (Est. FY'04) [154]	FY 2003, FSIS laboratories expanded capability: • To test for nontraditional microbial, chemical, and radiological threat agents. • Increased in surge capacity. • Has more than 7,600 inspectors and			

		veterinarians in more than 6000 Federal meat, poultry, and egg product plants, and at ports of entry, to prevent, detect, and respond to food-related emergencies. • Constructed a Bio-security Level-3 lab • Participated with HHS, EPA, DOE, and states to integrate the nation's laboratory infrastructure and surge capacity. • Over 60 labs in 27 states and 5 federal agencies have agreed to participate in Food Emergency Response Network (FERN)[155]	
Cooperative State Research, Education, and Extension Service (CSREES)	Extension/education/integrated activities—enacted for Homeland Security Mission:[156] • $31.6M (FY'03) • $39.2M (FY'04) • $66.3M (FY'05) Food safety[157] • $15M (Actual FY'03) • $13 (Est. FY'04) • $15M (Est. FY'05) Homeland security—Unified network of public agricultural institutions to identify and respond to high risk biological pathogens in the food and agricultural system:[158] • $8M (Est. FY'04) • $30M (Est. FY'05) The Agricultural Research, Extension, and Education reform Act of 1998 authorized the annual appropriation of $120 M for high priority research[159]		Critical plant/animal issues (funds to develop early intervention strategy to prevent, manage or eradicate new and emerging diseases, both plant and anaimal: • $1M (Actual, FY'03) • $1M (Es. FY'04) • $2M (Est. FY'05)

TABLE C-6 DHS Funding

Agency-Focus Area		Funding[160] FY 2003		FY 2004		FY 2005		FTEs
DHS		$31.2B		$36.5B		$40.2B		
SECURITY ENFORCEMENT & INVESTIGATIONS		$21.6B		$22.6B		$24.7B		
Bureau of Custom & Border Protection		$5.9 B		$5.9B		$6.2B		41,001
	(proposed for Agriculture Quarantine Program)		$407M					
PREPAREDNESS & RECOVERY		$5.2B		$5.5B		$7.4B		
Emergency Preparedness and Directorate Response (FEMA)					$5,493 M		$7,374M	2,511 FTEs (full time staff) 2,265 Disaster Staff
	NIMS						$7M	
	IMT						$8M	(20 FTEs)
	Biodefense						$2,528 M	
RESEARCH, DEVELOPMENT, TRAINING, ASSESSMENTS		$2.3B		$3.6B		$3.8B		
Science & Technology Directorate			$553M		$913M		$1,039M	
	Plum Island				$20M		$32.9M	
	Biosurveillance Initiative				$53M		$118M	
			$185M		$834M		$864M	
Information Analysis & Infrastructure Protection Directorate	Vulnerability Risk Assessment				$710 M		$702M	
	HSOC				$25M		$35M	
	Biosurveillance Initiative						$11M	

ter for Animal Health Modernization Project. This project will bring VS' NVSL and CVB together in one facility with the ARS' National Animal Disease Center, which will enhance collaboration. Subject to appropriations, construction will begin on the Low Containment (BSL-2) Large Animal Facility, the Consolidated Laboratory Building, Phase 2, and the balance of the infrastructure in FY2005, with completion of the entire project targeted for 2007.[161] When completed the National Centers for Animal Health would include nearly 1 million square feet of space with state-of-the-art capabilities for research and diagnosis.[162]

A pilot NAHLN involving 12 state/university diagnostic laboratories was funded through USDA in May 2002 for a 2-year period to develop capacity and surveillance programs.[163] A two-tiered funding structure is in place for the first 2 years, with tier 1 funded at $2M (laboratories in Calif., Colo., Ga., Tex., Wis.) and tier 2 at $750K (laboratories in Wash., Fla., N.Y., Iowa, Ariz., N.C., La.).[164] Though these funds provide critical pilot project start-up costs, they fall far short of developing a true national network that will effectively provide surveillance for zoonotic and foreign animal diseases, bioterrorist agents, and newly emergent diseases like West Nile virus. Federal funding to continue the pilot program and address these deficiencies is critically needed. Both additional start-up costs (estimated at $85M) and continuing funding (est. at $22M annually) have been requested in FY'05 to expand an enhanced, coordinated, and modernized NAHLN. Grants would be awarded directly to and coordinated by an accredited animal disease diagnostic laboratory when such a laboratory exists in the state, or in lieu of an accredited laboratory, to the primary animal disease diagnostic laboratory within the state.[165]

The NAHLN is in an early development stage so any money received for the state laboratories in FY 2004 will be concentrated on getting the 25 laboratories trained and provided with lab equipment, not adding an additional state. However, future targets reflect an increase of one new state each FY in 2005, FY 2006, and FY 2007.[166]

Currently, USDA has no BSL-4 facilities. BSL-4 facilities are required for research on pathogens that confer highly contagious, hot diseases, including the animal diseases bovine spongiform encephalopathy, as well as Hendrah and Nipah viruses.[167]

CDC-National Center for Infectious Diseases (NCID) whose mission is to prevent illness, disability, and death caused by infectious disease in the United States and around the world, conducts surveillance, epidemic investigations, epidemiological and laboratory research, training, and public education programs to develop, evaluate, and promote prevention and control strategies for infectious diseases. In October 2003, the center announced grants to build 11 new biodefense laboratories around the country, including 2 that will be authorized to study the most dangerous pathogens. Boston University Medical Center and the University of Texas Medical Branch at Galveston each will receive about $120M to build BSL-4 labs. BSL-4 labs can handle pathogens that pose a high risk of causing a life threatening diseases for which there is no vaccine or treatment, such as Ebola virus. In addition, the agency announced grants of $7M and $21M each for nine "Regional Biocontainment Laboratories," which include BSL-3 and BSL-2 space. Those facilities will be in Colorado State University, Fort Collins; Duke University, Durham, N.C.; Tulane University, New Orleans; the University of Alabama at Birmingham; the University of Chicago; the University of Medicine and Dentistry of New Jersey; the University of Missouri; the University of Pittsburgh; and the University of Tennessee.[168]

1.3.2 Veterinarian Capacity[169]

APHIS currently employs 2,053 veterinarians. Over half, 1,059, work for the Food Safety and Inspection Service (FSIS) and 531 work for APHIS. Of the 531 APHIS' veterinarians, 322 work in the field, a 20 percent decrease from the 404 field veterinarians in 1994. More significant is the comparison between today's resources and what was available in 1984. In 1984, when there was a large avian influenza outbreak in Pennsylvania, APHIS-VS had nearly 3,000 employees, in contrast to today's approximately 1,400. In contrast to shrinking resources, demand for investigation of suspected foreign animal diseases has risen from an average of 300 per year during the 1990s to 384 in 2000, 792 in 2001, and 837 in 2002. [170] This imbalance between demand and available human resources was characterized in the report on *Declining Infrastructure of Governmental Animal Health Professionals Puts American Agriculture at Risk* by Dr. Ron DeHaven, as follows:

> The current APHIS cadre of veterinarians and animal health professionals is clearly insufficient to handle the increased workload associated with trade obligations, emergencies, and already-apparent future demands. . . . APHIS faced two extensive outbreaks—END in California and last summer's avian influenza program in Virginia. . . .VS has detailed over half of its workforce to California, jeopardizing ongoing programs and leaving the United States vulnerable to any additional disease incursions. It is accurate to say, though very disturbing, that APHIS could not successfully respond to a significant foot and mouth disease (FMD) outbreak and continue to operate the END program in California.

To prepare for shortage of veterinarians, APHIS has developed the National Animal Health Reserve Corps to mobilize close to 300 private veterinarians from around the United States to assist locally during an emergency.[171]

2 DISEASES IN WILDLIFE

Managing animal health in wildlife is a daunting task. Reservoirs of infection in wild animals offer a constant threat to domestic livestock population and human health. Reducing the risk of transmission and spread may become an even larger factor in the future, with the potential for wildlife-based transmissible encephalopathies.[172]

2.1 THE DEPARTMENT OF THE INTERIOR, U.S. FISH AND WILDLIFE SERVICE (FWS)

FWS is responsible for the protection of wildlife from environmental hazards, safeguarding habitat for endangered species, and the inspection of wildlife shipments to ensure compliance with laws and treaties and detect illegal trade.[173] Generally, all wildlife imported into or exported from the United States for any purposes must be declared to the U.S. Fish and Wildlife Service and cleared prior to release by U.S. Customs and Border Protection. The U.S. Fish and Wildlife Service has a system of ports to allow for the import and export of wildlife, including parts and products.[174] Some wildlife inspection requires coordination with USDA APHIS, DOC's NMFS, INS, FDA and CDC.[175]

2.2 THE DEPARTMENT OF THE INTERIOR, BUREAU OF U.S. GEOLOGICAL SURVEY (USGS), BIOLOGICAL RESOURCES DIVISION (BRD), THE NATIONAL WILDLIFE HEALTH CENTER (NWHC)

NWHC is one of the 18 science and technology centers in the BRD of the USGS, a bureau of the DOI, located in Madison, Wis. The NWHC was established in 1975 as a biomedical laboratory dedicated to assessing the impact of disease on wildlife and to identifying the role of various pathogens contributing to wildlife losses.[176] The center provides a multi-disciplinary, integrated program of disease diagnosis, field investigation and disease management, research, and training, and maintains extensive databases on disease findings in animals and on wildlife mortality events.

NWHC has over 70 scientists (specialists in such fields as wildlife ecology, epidemiology, veterinary medicine, pathology, virology, bacteriology, parasitology, chemistry, biometry, and population ecology) and support personnel. Center field personnel respond to catastrophic events, such as major die-offs, that threaten the health of wildlife populations. The NWHC has specialized biological containment facilities that allow investigation of infectious diseases affecting a broad spectrum of wildlife, such as amphibians, eagles, sea turtles, sea otters, migratory birds, wolves, large mammals, and other species. National wildlife refuge personnel, law enforcement agents, state conservation agency biologists, university-affiliated scientists, and others send wildlife carcasses and tissue samples to the NWHC for diagnostic examination.[177,178,179]

The NWHC is an international focal point for research, information, and exchange of information on the study of wildlife health and disease. The center's researches on zoonotic diseases concentrate on better understanding the ecological relationships among free-ranging wildlife, domestic animals, and public health concerns. This understanding is fundamental for developing effective disease prevention and control strategies. Other center research is directed toward developing enhanced technology for disease detection and diagnosis and toward developing biologics to protect animals against infection.[178,179]

NWHC is currently monitoring the outbreak in Western Europe and other parts of the world and gathering information from numerous sources. Specialists are interacting with the USDA, and working with the U.S. DOI land management and conservation agencies to provide information on disease status and risks and assist in developing FMD prevention and contingency plans.[180] Congress requested that USDA and DOI work together to create a national plan to assist the states and tribes in addressing CWD in both farmed and wild animals. Budget and implementation plan has been developed. Implementation of this plan is proceeding as budgets allow.[181]

With available specialized containment facilities, NWCH is also providing diagnostic support and research results to federal, state, and local wildlife agencies, as well as public health departments that are utilizing dead wild birds as sentinel, for detecting the West Nile virus (WNV). USGS is working with CDC to learn the current geographic extent of WNV. Scientist at USGS Geographic Science Branch are providing CDC and public health agencies with real-time geographic information on land-use and land-cover data, roads, and hydrography in areas where the virus is active. These data are used to identity bird and mosquito habitat for placement of mosquito traps.[182]

2.3 THE DEPARTMENT OF INTERIOR, USGS, NATIONAL BIOLOGICAL INFORMATION INFRASTRUCTURE (NBII) PROGRAMS

The NBII is a broad collaborative program to provide increased access to data and information on the national's biological resources. This web-based, biological information system draws together vast amounts of scientific data and provides information via the Web. NBII partners include fish and wildlife agencies and including NOAA, EPA, National Science Foundation, Oak Ridge National Laboratory (ONL), USDA-ARS, USDA Forest Service, U.S. Fish and Wildlife Service.[183,184]

NBII was created in 1993, based on the recommendation of a special panel convened by the National Research Council to examine critical national biological resource issues. In 1998, the need for the NBII was reaffirmed by a team of internationally renowned scientists who also recommend the creation of NBII "nodes" as focal points for various biological and regional issues. The NBII Program initiated 10 nodes in FY 2001 and began new prototype in FY 20002. The Fisheries and Aquatic Resource node provides access to fisheries information resources from across the world.[185] The Wildlife Disease Information Node is to develop collaborative national database of wildlife mortality events to facilitate tracking and study of emerging wildlife diseases such as WNV and CWD.[186]

2.3.1 THE NBII WILDLIFE DISEASE INFORMATION NODE (WDIN)[187]

The WDIN provides access to near real-time data on wildlife mortality events and other critical related information. The major objectives of the Wildlife Disease Information Node include:

- Documenting the prevalence and spread of wildlife diseases at the most discrete spatial and temporal levels possible via a nationwide web-based reporting system.
- Maintaining current databases on wildlife mortality events and other critical information.
- Providing Web access to wildlife and zoonotic disease information for management, research, epidemiological, and educational purposes.

• Providing Web access to the general public for educational purposes and to disseminate information on the importance of wildlife and zoonotic diseases and related ecosystem and community effects.

• Developing partnerships to share wildlife mortality and other critical information in a distributed fashion and in a secure, partner-based data system.

In FY 2004, the major WDIN undertaking in partnership with NWHC was the development of a robust Web-based CWD national clearinghouse that can accommodate contributed testing results as well as research, monitoring, and surveillance data from state, federal, and tribal agencies as well as other organizations doing CWD work into a common database scheme. Mechanisms by which data can be queried, analyzed, and visualized to make CWD data and information accessible to all parties dealing with CWD issues are being established. Wild and captive cervid data would be included.[188]

A prototype effort is also being funded by the National Science Foundation and the Intelligence Technology Innovation Center, which is aimed at developing scalable technologies and related standards and protocols needed for a full implementation of a national infectious disease information infrastructure for human, plant, and animal (domestic and wild) diseases. The interdisciplinary team consists of the following groups for this prototype effort:

• Artificial Intelligence Laboratory at the University of Arizona.

• The New York State Department of Health and its partner Health Research Inc.

• The California State Department of Health Services and its partner PHFE Management Solutions.

• The U.S. Geological Survey's National Biological Information Infrastructure and the National Wildlife Health Center.

The two diseases selected for this prototype are West Nile virus and botulism because of their significant public health and homeland security implications. After an extensive 4-month research and system development efforts, the University of Arizona completed a research prototype called the WNV-BOT Portal System, which provides integrated, Web-enabled access to a variety of distributed data sources related to WNV and botulism. It also provides advanced information visualization capabilities as well as predictive modeling support.[188]

2.4 USDA-APHIS-VS WILDLIFE SERVICE (WS)[189]

WS provides expertise to resolve wildlife conflicts that threaten livestock and protect livestock from wildlife-borne diseases such as CWD. WS works with state counterparts to monitor wildlife diseases like rabies and has cooperative agreements with state wildlife agencies to fund surveillance and management.

The National Wildlife Research Center (NWRC) in Fort Collins, Colo., functions as WS' research arm. It is the only federal research facility devoted exclusively to resolving conflicts between people and wildlife. WS conducts research projects to study the interaction of deer and cattle, coyotes as reservoirs of tuberculosis, and CWD in elk.

The WS Aquatic Animal Health (AAH) program provides diagnostic assistance to aquaculture producers experiencing health problems with their products. VS works with U.S. Department of the Interior's U.S. Fish and Wildlife Service (FWS), the U.S. Department of Commerce's National Marine Fisheries Service, and members of the Joint Subcommittee on Aquaculture to disseminate information and outline agency roles to best meet the aquaculture industry's animal health needs and develop aquatic animal health monitoring and surveillance program.

3 FISHERIES

3.1 DEPARTMENT OF COMMERCE, NATIONAL OCEANIC AND ATMOSPHERIC ADMINISTRATION, NATIONAL MARINE FISHERIES SERVICE (NMFS)

NOAA oversees fisheries management in the United States and, through the 1946 Agriculture Marketing Act, provides a voluntary inspection service to the industry. The NOAA Seafood Inspection Program offers a variety of professional inspection services, which assure compliance with all applicable food regulations. These services are available nationwide, at all types of establishments such as vessels, processing plants, and retail facilities. All edible product forms ranging from whole fish to formulated products and fishmeal products used for animal foods are eligible for inspection and certification.[190]

Aquaculture: Global aquaculture now produces more than 31 million metric tons of farm products (fish, crustaceans, and mollusks) annually, which have a value of some $38 billion. The United States is eighth among leading aquaculture producers worldwide, with annual market share value approaching $1 billion.[191]

Statutory authority: The National Aquaculture Act of 1980, as amended (16 U.S.C. 22801 et seq.) allows the development of a U.S. aquaculture industry. The Act established the Joint Sub-Committee on Aquaculture as coordination group for the federal government activities relating to aquaculture, and charged JSA with development of a National Aquaculture Development Plan. Amendments to the Act in 1985 designated the Secretary of Agriculture as the permanent chair of the JSA. The secretaries of Agriculture, Commerce, and Interior make up the Executive

Committee.[192] These agencies are among those listed as resources to aquaculture programs and services within the federal government by the JSA:[193]

- U.S. Department of Agriculture: ARS, CSREES, Regional Aquaculture Centers, Farm Service Agencies, AMS, NASS, FAS, APHIS, Federal Crop Insurance Information, National Agriculture Library, Alternative Farming Systems Information Center, Current Research Information System
- U.S. Department of Commerce: NMFS, National Sea Grant College Program, Economics and Statistic Administration, National Weather Service, National Environmental Satellite Data and Information Service
- U.S. Department of Interior: U.S. Fish and Wildlife Service, U.S. Geological Survey
- U.S. DHHS: FDA-CVM, FDA-CFSAN
- National Science Foundation
- EPA
- Tennessee Valley Authority (TVA)
- U.S. Agency for International Development (USAID)

4 FOOD SAFETY

Four federal agencies share primary responsibility for federal food safety. The largest of these, the USDA-FSIS, regulates meat and poultry through continuous inspection of processing operations and review and approval of product labels. The EPA Office of Pesticide Program (OPP) register pesticides and sets tolerances that are enforced by FDA and FSIS. Finally, the CDC is the federal government's primary clearinghouse for disease morbidity and mortality surveillance data and its chief resource for epidemiological investigations.[194] See Table C-7 for food safety responsibilities for selected food products.

4.1 USDA-FSIS

FSIS inspects most meat, poultry, and processed eggs sold for human consumption for safety, wholesomeness, and proper labeling. The Federal Meat Inspection Act of 1906, as amended [21USC 601 et seq.] requires USDA to inspect all cattle, sheep, swine, goats, and horses brought into any plant to be slaughtered and processed into products for human consumptions. The Egg Products Inspection Act, as amended [21USC 1031 et seq.] is the authority under which FSIS ensure the safety of egg products. For bioterrorism preparedness, FSIS' Food Biosecurity Action Team (F-BAT) has placed the agency's 7,600 inspectors on high alert to look for antemortem and postmortem irregularities in meat animals and poultry, and has conducted mock exercises to respond in emergency situations. The Food Threat Preparedness Network (PrepNet) is a joint FSIS/FDA group that works on threat prevention and emergency.[195]

TABLE C-7 Food Safety Responsibilities for Selected Food Products[204]

Food	Regulators	Comments
Eggs	FDA, AMS, FSIS, APHIS	FDA has lead jurisdiction over shell eggs; FSIS continuously inspects egg products. AMS operates a voluntary grading program. APHIS monitors animal health.
Meat and poultry	FSIS, FDA	FSIS inspects meat during processing. FDA holds regulatory authority once meat leaves the slaughtering or manufacturing plant.
Processed foods	FDA	FDA is responsible for most nonmeat products.
Seafood	FDA, NMFS	FDA oversees seafood safety generally. NMFS run a voluntary inspection service.

4.2 FDA

The FDA regulates 80 percent of the nation's food supply, except for meat, poultry and certain egg products. Through its Center for Food Safety and Applied Nutrition (CFSAN), FDA monitors the safety and labeling of most nonmeat and processed foods and licenses food-use chemicals other than pesticides. CFSAN has the authority to regulate food producers and distributors involved in interstate commerce and to issue recommendations on food safety issues, including foods and cosmetics using bovine ingredients.[196] FDA also operates an oversight compliance program for fishery products under which responsibility for the product's safety, wholesomeness, identity, and economic integrity rests with the processor or importer, who must comply with regulations promulgated under the Federal Food, Drug and cosmetic (FD&C) Act, as amended, and the Fair Packaging and Labeling Act (FPLA). In addition, FDA operates the Low-Acid Canned Food (LACF) Program, which is based on the hazard analysis critical control point (HACCP) concept and is focused on thermally processed, commercially sterile foods, including seafood such as canned tuna and salmon.[197]

Title III of the Bioterrorism Act provide the HHS Secretary with new authorities to protect the nation's food supply against the threat of intentional contamination and other food-related emergencies. The interim final rule promulgated under this act requires domestic and foreign facilities that manufacture or process, pack, or hold food for human or animal consumption in the United State to register with FDA. FDA expects up to 420,000 facilities to register under this requirement.[198]

The FDA-Center for Veterinary Medicine (FDA-CVM) ensures the safety, efficacy and quality of drugs used in animals, including animal feed and companion animals, animal food and feed, and medical devices used on animals. CVM regulates all animal drugs and feed and works to increase the availability of products to sustain the health, relieve the suffering, and increase the productivity of all farm animals. CVM's current top three priorities are to prevent BSE, counter the risk of food associated with antibiotic resistance in humans, and ensure safe food derived from genetically modified animals.[199]

CVM monitors and establishes standards for feed contaminants, approves safe food additives, and manages the FDA's medicated feed and pet food programs. Office of Surveillance and Compliance monitors marketed animal drugs, food additives, and veterinary devices. Also involved in these activities are the USDA, EPA, and other state and other federal agencies.[200]

CVM's specifics activities in a BSE emergency response include[201]:

- Collaborating with public health agencies (CDC, HHS, and USDA) and with states, regarding feed contaminant, tissue residue programs, and other monitoring programs for meat and poultry involving a BSE emergency
- Providing information regarding manufacturer's GMP compliance and other relevant animal drug quality issues
- Providing advice in the assessment of animal drugs or feed products involving a BSE emergency

4.3 CDC

In the last decade, the CDC established more than 10 surveillance systems to identify and track the source of outbreaks of foodborne illnesses and to assist regulatory agencies in their food safety activities. Consequently, the agency now has separate surveillance systems to track botulism, Creutzfeldt-Jacob disease (human form of mad cow disease), *E. coli* O157:H7, *Giardia*, *Salmonella*, and *Salmonella enteritidis*, viral hepatitis, trichinellosis, typhoid fever, and *Vibrio* infections in foods.[202] CDC's surveillance systems for the most part depend on reporting capabilities of local- and state-level health and agriculture officials. Since 9/11/2001, the agency has been training these officials and laboratory technicians to recognize hazards in foods. It has also begun to refurbish public health laboratories in most states to increase the capacity of these facilities to quickly identify the act of terrorism.[203]

In addition to the four major organizations, there are a number of other federal agencies with ancillary or supporting roles in the government's regulatory programs to ensure food safety and counterterrorism efforts. These partners include: the USDA's Agricultural Marketing Service; the USDA's Grain Inspection, Packers, and Stockyards Administration (GIPSA); the USDA's Office of Risk Assessment and Cost-Benefit Analysis; the USDA's Agricultural Research Service (ARS); the USDA's Animal and Plant Inspection Service (APHIS); the USDA's Cooperative State Research, Education, and Extension Service (CSREES); the USDA's Economic Research Service (ERS); USDA-Foreign Agricultural Service, USDA-ARS, USDA-Food and Nutrition Service, Department of Commerce's (DOC) National Oceanic and Atmospheric Administration (NOAA), the Treasury Department's Bureau of Alcohol, Tobacco and Firearms (ATF); Department of State, the Federal Trade Commission (FTC); Custom and Border Protection (CBP) in DHS; Department of Army Veterinary Services Activity, Department of Treasury's Alcohol and Tobacco Tax and Trade Bureau (TTB), the FBI and CIA.[205,206]

There is an increased participation by veterinary diagnostic labs in public health programs, as evident in these programs:[207] ELEXNET—an integrated web-based data exchange system for food testing information that allows multiple agencies engaged in food activities to compare and communicate and coordinate findings of laboratory analyses;[208] FERN (food emergency response network)—to increase surge capacity. FDA and USDA-FSIS are working with CDC to expand the LRN to include a substantial number of counterterrorism laboratories capable of analyzing foods for agents of concern. As of November 2003, there are 63 labs representing 27 states expressing interest in participation in FERN), CELDAR (CA DHS and CAHFS).

5 Endnotes

[1]USDA-ARS Program Summary: Program Rationale. (http://www.ars.usda.gov/research/programs/programs.htm?np_code=103?docid=820; accessed 3/1/04)

[2]DeHaven, Ron. A Report on "Declining Infrastructure of Governmental Animal Health Professionals Puts American Agriculture at Risk." August 2003.

[3]USDA-ARS, Program Summary: Program Rationale. (http://www.ars.usda.gov/research/programs/programs.htm?np_code=103&docid=820; accessed 3/1/04)

[4]USDA-ARS Program Summary: Program Rationale. (http://www.ars.usda.gov/research/programs/programs.htm?np_code=103?docid=820; accessed 3/1/04)

[5]USDA-APHIS-VS: ESF 11b. Animal production. National Animal Health Emergency Response Plan for an Outbreak of Foot-And-Mouth Disease or Other Highly Contagious Animal Diseases (February 11, 2004)

[6]US Congress, 2001, as cited in NAS-NRC Countering Agricultural Bioterrorism, 2003

[7]USDA-Veterinary Services Strategic Plan, FY 2004 to FY 2008, Updated February 2004

[8]http://www.aphis.usda/gov/lpa/about/strategic_plan/archive/strategic_plan_00-05/strategic_mission.htm; accessed 2/27/04

[9] CDC Fact Sheet, NCID, Division of Global Migration and Quarantine, Importation of Pets, Other Animals, and Animal Products into the United States. (www.cdc.gov/ncidod/q/animal.htm; accessed 3/18/04)

[10]Creekmore,L., USDA, APHIS, VS, NCAHP, Eradication and Surveillance Team, Power Point Slide Presentation.

[11]Grannis, J., Center for Emerging Issues, USDA-VS, CEAH, Power Point Slide Presentation.

[12]OMB Budget Documents, Department of Agriculture Part Assessment. (http://www.whitehouse.gov/omb/budget/fy2005/pma/agriculture.pdf; accessed 4/12/04)

[13]Grannis, J., Center for Emerging Issues, USDA-VS, CEAH, Power Point Slide Presentation.

[14]USDA-APHIS-VS: ESF 11b Animal Production. National Animal Health Emergency Response Plan for an Outbreak of Foot-And-Mouth Diseases or Other Highly Contagious Animal Diseases (2/11/2004)

[15]OMB Budget Documents, Department of Agriculture Part Assessment. (http://www.whitehouse.gov/omb/budget/fy2005/pma/agriculture.pdf; accessed 4/12/04)

[16]ARS, FY2000 and 2001 Annual Performance Plan

[17]CFDA, 10.207 Animal Health and Disease Research. (http:www.cfda.gov/public/viewprog.asp?progid=38; accessed 4/8/04)

[18]CRS, Issue brief for Congress, Food Safety and Protection Issues in the 107th Congress, Updated August 28, 2002. No.1B00099.

[19]Crawford, Lester, Statement before the Committee on Government Affairs, U.S. Senate, November 19, 2003.

[20]CDC Fact Sheet, NCID-Division of Global Migration and Quarantine (http://wwwcdc.gov/ncidod/dq/mission.htm; accessed 4/6/04)

[21]CDC Fact Sheet, Office of Health and Safety, Etiologic Agent Import Permit Program. (http://www.cdc.gov/od/ohs/biosfty/imprtper.htm; accessed 4/6/04)

[22]CDC Fact Sheet, NCID, Division of Global Migration and Quarantine, Importation of Pets, Other Animals, and Animal Products into the United States (http://www.cdc.gov/ncidod/dq/animal.htm; accessed 3/18/04)

[23]USDA-VS Strategic Plan FY 2004 to FY 2008, Updated February 2004.

[24]APHIS Fact Sheet, The Animal and Plant Health Inspection Service and Department of Homeland Security: Working Together to Protect Agriculture. May 2003.

[25]OMB FY2005 Budgets. (http://www.whitehouse.gov/omb/budget/fy2005/pdf/ap_cd_rom/homeland.pdf)

[26]President George W. Bush, Securing the Homeland, Strengthening the Nation

[27]Creekmore, L., USDA, APHIS, VS, NCAHP, Eradication and Surveillance Team, Power Point Slide Presentation.

[28]ARS, FY 2000 and 2001 Annual Performance Plans.

[29]USDA-Agricultural Research Service, About ARS. (http://www.ars.usda.gov/aboutus/accessed 2/27/04)

[30]USDA-ARS, National Programs. (http://www.ars.usda.gov/research/programs.html; accessed 2/27/04)

[31]USDA, ARS National Program, Program Summary: Program Component Definitions. (http://www.ars.usda.gov/research/programs/program.htm?np_code=303&docid=795; accessed 3/1/04)

[32]USDA-National Animal Disease Center. (http://www.nadc.ars.usda.gov/about/Mission/bmission.asp; accessed 3/1/04)

[33]CFDA, 10.207 Animal Health and Disease Research. (http:www.cfda.gov/public/viewprog.asp?progid=38; accessed 4/8/04)

[34]CSREES, NRI. (http://www.reeusda.gov/nri/nriinfo/about.htm; accessed 4/13/04)

[35]Dr. J. Annelli, APHIS-VS-EP, Personal Communication, April 2004.

[36]NAS-NRC, Countering Agricultural Bioterrorism, The National Academies Press, 2003.

[37]APHIS Fact Sheet, The Animal and Plant Health Inspection Service and Department of Homeland Security: Working Together to Protect Agriculture. May 2003

[38]APHIS Fact Sheet, The Animal and Plant Health Inspection Service and Department of Homeland Security: Working Together to Protect Agriculture. May 2003.

[39]Dr. Annelli, APHIS-VS-EP, Personal Communication, April 2004.

[40]USDA-VS Strategic Plan FY 2004 to FY 2008 Updated February 2004.

[41]DHS Organization - Dept Components. (http://www.dhs.gov/dhspublic/display?theme=9&content=2973; accessed 3/1/04)

[42]Testimony of Commissioner Robert C. Bonner, US CBP Before the National Commission on Terrorist Attacks Upon the US, January 26, 2004.

[43]FCBF News Flash - CBP Agriculture Specialist Fact Sheet - 9/1/01. (http://www.fcbf.com/NewsFlashDetail.asp?NewsId=47; accessed 4/6/04)

[44]FCBF News Flash - CBP Agriculture Specialist Fact Sheet - 9/1/01. (http://www.fcbf.com/NewsFlashDetail.asp?NewsId=47; accessed 4/6/04)

[45]USAHA 2003 Resolution No. 21 from the 2003 USAHA Annual Meeting Oct. 15, 2003. (http://www.usaha.org/resolutions/reso03/res-2103.html; Accessed 4/6/2004)

[46]CBP today, March 2003. (March/ag.xml" http://www.cbp.gov/xp/CustomsToday/2003 March/ag.xml; accessed 3/31/2004)

[47]US CBP website press release Jan. 14, 2004. (http://www.customs.gov/xp/cgov/newsroom/press_releases/0012004/01142004_4.xml; accessed 4/2/2004

[48]No retraining for Agricultural Inspectors in Border Agency Plan. (10/30/2003) (http://www.fass.org/fasstrack/news_item.asp?news_id=1646; accessed 4/6/2004)

[49]Dr. Ahmad, Veterinarian in Charge at JFK (718-553-1727). Telephone communication, (4/6/05).

[50]DHS Protecting Against Agricultural Terrorism, (http://www.dhs.gov/dhspublic/display?theme=43&content=3117; accessed 4/2/04)

[51]CDC Fact Sheet, Office of Health and Safety, Etiologic Agent Import Permit Program. http://www.cdc.gov/od/ohs/biosfty/imprtper.htm; accessed 4/6/04)

[52]CDC Fact Sheet, NCID, Division of Global Migration and Quarantine, Importation of Pets, Other Animals, and Animal Products into the United States. (http://www.cdc.gov/ncidod/dq/animal.htm; accessed 3/18/04)

[53]CDC Fact Sheet, NCID, Division of Global Migration and Quarantine, History of Quarantine. (http://www.cdc.gov/ncidod/dq/history.htm)

[54]NCIE Fact Sheet. (www.aphis.usda.gov/vs/ncie/; accessed 3/1/04)

[55]Statement of Dr. Charles Lambert, Senate Committee on Government Affairs. "Agroterrorism: The Threat to America's Breakfast." November 19, 2003. (http:/govt-aff.senate.gov/index.cfm?Fuseaction=Hearings.Testimony&HearingID=127& WitnessID =482; accessed 4/2/2004)

[56]NAS-NRC, Countering Agricultural Bioterrorism, National Academy Press, 2003.

[57]NAS-NRC, Countering Agricultural Bioterrorism, National Academy Press, 2003

[58]USDA VS Strategic Plan FY 2004 to FY 2008, updated February 2004

[59]Accord, Bobby, Statement before the House Subcommittee on Agriculture, Rural Development, Food and Drug Administration, and Related Agencies.

[60]NAS-NRC, Countering Agricultural Bioterrorism, The National Academies Press, 2003.

[61]Dr. J. Annelli, APHIS-VS-EP, Personal Communication, April 2004.

[62]Dr. J. Annelli, APHIS-VS-EP, Personal Communication, April 2004.

[63]HSPD9, January 30, 2004. (http://www.whitehouse.gove/news/releases/2004/02/20040203-2.html; accessed 3/32/04)

[64]APHIS Fact Sheet, Veterinary Services, July 2003.

[65]APHIS Fact Sheet, Veterinary Services, July 2003.

[66]Accord, Bobby, Statement before the House Subcommittee on Agriculture, Rural Development, Food and Drug Administration, and Related Agencies.

[67]Bruntz, Stan, APHIS-VS-CEAH, Presentation on the National Animal Health Reporting System

[68]Bruntz, Stan, APHIS-VS-CEAH, Presentation on the National Animal Health Reporting System

[69]Dr. J. Annelli , APHIS-VS-EP, Personal Communication, April 2004.

[70]APHIS Fact Sheet, Veterinary Services, July 2003.

[71]Dr. J. Annelli, APHIS-VS-EP, Personal Communication, April 2004.

[72]USAHA, Report of the Joint USAHA/AAVLD Committee on Animal Health Informa-

tion Systems, 2003 Reports. (Http://www.usaha.org/reports/reports03/r03ahis.html; accessed 4/12/04)

[73]APHIS Fact Sheet, Veterinary Services, July 2003.

[74]APHIS-VS Strategic Plan FY 2004 to FY 2008, Updated February 2004.

[75]USDA-VS Strategic Plan FY 2004 to FY 2008, Updated February 2004.

[76]USDA-VS Strategic Plan FY 2004 to FY 2008, Updated February 2004.

[77]USAIP, the United States Animal Identification Plan, January 2004.

[78]NAHEMS Steering Committee. Standard for Animal Health Emergency Management Systems. January 2000.

[79]NAS-NRC, Countering Agricultural Bioterrorism, The National Academies Press, 2003.

[80]APHIS Fact Sheet, Veterinary Service, July 2003.

[81]APHIS-VS Strategic Plan FY 2004 to FY 2008, updated February 2004.

[82]Roy, Alma, Louisiana Vet Med Diag Lab, School of Vet Med, LSU, Presentation on Animal Diagnostic Capabilities.

[83]NAS-NRC, Countering Agricultural Bioterrorism, The National Academies Press, 2003.

[84]USDA-VS Strategic Plan FY 2004 to FY 2008. Updated February 2004.

[85]USDA Fact Sheet, National Animal Health Network, May 30, 2003.

[86]APHIS Fact Sheet, Veterinary Service, July 2003.

[87]USAHA, Report of the Joint USAHA/AAVLD Committee on Animal Health Information Systems. 2002 Committee Report. (http://www.usaha.org/reports/report02/r02ahis.html; accessed 4/12/04)

[88]APHIS Fact Sheet, Veterinary Service, July 2003.

[89]APHIS Fact Sheet, Veterinary Service, July 2003.

[90]USAHA, Report of the Joint USAHA/AAVLD Committee on Animal Health Information Systems. 2002 Committee Reports. (http://www.usaha.org/reports/reports02/r02ahis.html)

[91]USAHA, Report of the Joint USAHA/AAVLD Committee on Animal Health Information Systems. 2002 Committee Reports. (http://www.usaha.org/reports/report02/r02ahis.html; accessed 4/12/04)

[92]USDA VS Strategic Plan FY2004 to FY 2008, Updated February 2004.

[93]USDA VS Strategic Plan FY2004 to FY 2008, Updated February 2004.

[94]Gilchrist, Mary, "The progress, priorities and concerns of public health laboratories." Forum on Emerging Infections, Biological Threats and Terrorism, Institute of Medicine, November 28, 2001.

[95]CDC Fact Sheet, Laboratory Preparedness for Emergencies. February 4, 2004.

[96]CDC Fact Sheet, Laboratory Preparedness for Emergencies. February 4, 2004.

[97]CDC Fact Sheet, Laboratory Preparedness for Emergencies, February 4, 2004.

[98]CDC, National Center for Infectious Diseases. (http://www.cdc.gov/ncidod/about.htm; accessed 3/18/04)

[99]CDC Fact Sheet, Laboratory Preparedness for Emergencies. February 4, 2004.

[100]NAS-NRC, Countering Agricultural Bioterrorism, The National Academies Press, 2003.

[101]OMB Budget Documents. (http://www.whitehouse.gov/omb/budget/fy2005/appendix.html)

[102]CFDA, 10.207 Animal Health and Disease Research. (http:www.cfda.gov/public/viewprog.asp?progid=38; accessed 4/8/04

[103]CSREES, FY2005 President's FY2005 Budget Proposal — Advancing Knowledge for the Food and Agricultural System, Overview. Feb 2004.

[104]CSREES, NRI. (http://www.reeusda.gov/nri/nriinfo/about.htm; accessed 4/13/04)

[105]CSREES, FY2005 President's FY2005 Budget Proposal — Advancing Knowledge for the Food and Agricultural System, Overview. Feb 2004.

[106] USDA Fact sheet, Homeland Security Efforts, June 2003.

[107]USDA-ARS National Programs. (http://www.ars.usda.gov/research/programs.htm; accessed 2/27/04)

[108]USDA-ARS Action Plan. (http://www.ars.usda.gov/research/programs/programs.htm?np_code=103?docid=820; accessed 3/1/04)

[109]USDA-ARS Action Plan. (http://www.ars.usda.gov/research/programs/programs.htm?np_code=103?docid=820; accessed 3/1/04)

[110]Iowa State University, USDA and International Resources. (http://www.vetmed.iastate.edu/academics/international/resources.html; accessed 4/12/04)

[111]USDA-ARS Action Plan. (http://www.ars.usda.gov/research/programs/programs.htm?np_code=103?docid=820; accessed 3/1/04)

[112]USDA-ARS Action Plan, Epidemiology of Diseases. (http://www.ars.usda.gov/research/programs/programs.htm?np_code=103?docid=820; accessed 3/1/04)

[113]DHS Research and Technology. (http://www.dhs.gov/dhspublic/display?theme=27&content=937; accessed 3/1/04)

[114]Albright, Penrose, Assistant Secretary for Science and Technology, Department of Homeland Security, Testimony before the U.S. Senate Committee on Governmental Affairs, November 19, 2003.

[115]DHS-Homeland Security Centers Program. (http://www.cals.wisc.edu/research/hlsecurity.html; accessed 4/5/04)

[116]Merril, R.A., and J.K. Francer, Organizing Federal Food Safety Regulation. Seton Hall Law Review, No.1,Vol.31, 2000.

[117]USAMRIID Fact Sheet. (http://www.usarmiid.army.mil/programs/index.html; accessed 3/8/04)

[118]McNair Paper 65. Chapter 4, Recommendations and Conclusions (http//www.ndu.edu/inss/McNair/mcnair65/08_ch04.htm; accessed 4/12/04)

[119]CDC Continuation Guidance for Cooperative Agreement on Public Health Preparedness and Response for Bioterrorism—Budget Year Four, Program Announcement 99051. May 2, 2003.

[120]Homeland Security Presidential Directive/HSPD-5.

[121]Accord, Bobby, Statement before the House Subcommittee on Agriculture, Rural Development, Food and Drug Administration, and Related Agencies.

[122]NAHEMS Strategic Plan for years 2000 through 2005. (http://www.usaha.org/NAHEMS/Strategy.html; accessed 3/1/04)

[123]USDA-APHIS-VS: ESF 11b. Animal Production. National Animal Health Emergency Response Plan for An Outbreak of Foot-and-Mouth Disease or Other Highly Contagious Animal Diseases (February 11, 2004).

[124]USDA-APHIS-VS: ESF 11b. Animal Production. National Animal Health Emergency Response Plan for An Outbreak of Foot-and-Mouth Disease or Other Highly Contagious Animal Diseases (February 11, 2004)

[125]McNair Paper 65. Chapter 4, Recommendations and Conclusions (http//www.ndu.edu/inss/McNair/mcnair65/08_ch04.htm; accessed 4/12/04)

[126]USDA-APHIS-VS: ESF 11bAnimal production. National Animal Health Emergency Response Plan For An Outbreak of Foot-And-Mouth Disease or Other Highly Contagious Animal Diseases (February 11, 2004).

[127]Dr. J. Annelli, APHIS-VS-EP, Personal Communication, April 2004.

[128]USDA-APHIS-VS: ESF 11b Animal production. National Animal Health Emergency Response Plan For an Outbreak of Foot-and-Mouth Disease or Other Highly Contagious Animal Diseases (February 11, 2004).

[129]Accord, Bobby, Administrator, APHIS, Statement before the House Subcommittee on Agriculture, Rural Development, Food and Drug Administration, and Related Agencies.

[130]APHIS Veterinary Services Fact Sheet, March 2003.

[131]APHIS Veterinary Services Fact Sheet, The New APHIS Emergency Operations Center, March 2003.

[132]USDA-APHIS-VS: ESF 11b Animal Production. National Animal Health Emergency Response Plan For an Outbreak of Foot-and-Mouth Disease or Other Highly Contagious Animal Diseases (February 11, 2004).

[133]USDA-APHIS-VS: ESF 11b Animal Production. National Animal Health Emergency Response Plan for An Outbreak of Foot-and-Mouth Disease or Other Highly Contagious Animal Diseases (February 11, 2004).

[134]APHIS Fact Sheet, Veterinary Services, July 2003.

[135]Dr. J. Annelli, APHIS-VS-EP, Personal Communication, April 2004.

[136]Accord, Bobby, Administrator, APHIS, Statement before the House Subcommittee on Agriculture, Rural Development, Food and Drug Administration, and Related Agencies.

[137]Accord, Bobby, Administrator, APHIS, Statement before the House Subcommittee on Agriculture, Rural Development, Food and Drug Administration, and Related Agencies.

[138]Iowa State University, USDA and International Resources. (http://www.vetmed.iastate.edu/academics/international/resources.html; accessed 4/12/04)

[139]Iowa State University, USDA and International Resources. (http://www.vetmed.iastate.edu/academics/international/resources.html; accessed 4/12/04)

[140]Statement of Dr. Charles Lambert, Senate Committee on Government Affairs. "Agroterrorism: The Threat to America's Breakfast." November 19, 2003. (http:/govt-aff.senate.gov/index.cfm?Fuseaction=Hearings. Testimony&HearingID=127&WitnessID=482; accessed 4/2/2004)

[141]David Kinker, National Animal Health Laboratory Network, Presented to the Secretary's Committee on Foreign Animal and Poultry Disease, Riverdale, MD, February 4, 2004; Committee on Foreign Animal Diseases of the United States Animal Health Association, Foreign Animal Diseases, 1998 Edition, Richmond, VA: USAHA, 1998. (Cited in House Democrats statement 2004)

[142]House Democrats, 2004 House Democrats Statement: Protecting the Food Supply. (www.houselgov/hsc/democrats/pdf/press/protectingthefoodsupply.pdf, Accessed 4/5/04)

[143]OMB Budget Documents. (http://www.whitehouse.gov/omb/budget/fy2005/appendix. html)

[144]OMB Budget Documents. (http://www.whitehouse.gov/omb/budget/fy2005/appendix. html)

[145]OMB Budget Documents. (http://www.whitehouse.gov/omb/budget/fy2005/appendix. html)

[146]OMB Budget Documents. (http://www.whitehouse.gov/omb/budget/fy2005/appendix. html)

[147]OMB Budget Documents. (http://www.whitehouse.gov/omb/budget/fy2005/appendix. html)

[148]OMB Budget Documents. (http://www.whitehouse.gov/omb/budget/fy2005/appendix.html)

[149]OMB Budget Documents. (http://www.whitehouse.gov/omb/budget/fy2005/appendix. html)

[150]OMB Budget Documents. (http://www.whitehouse.gov/omb/budget/fy2005/appendix. html)

[151]Statement of Dr. Charles Lambert, Senate Committee on Government Affairs. "Agroterrorism: The Threat to America's Breakfast." November 19, 2003. (http:/govt-aff.senate.gov/index.cfm?Fuseaction=Hearings.Testimony&HearingID=127&WitnessID=482; accessed 4/2/2004)

[152]OMB Budget Documents. (http://www.whitehouse.gov/omb/budget/fy2005/appendix. html)

[153]OMB Budget Documents. (http://www.whitehouse.gov/omb/budget/fy2005/appendix. html)

[154]OMB Budget Documents. (http://www.whitehouse.gov/omb/budget/fy2005/appendix. html)

[155]Statement of Dr. Charles Lambert, Senate Committee on Government Affairs. "Agroterrorism: The Threat to America's Breakfast." November 19, 2003. (http:/govt-aff.senate. gov/index.cfm?Fuseaction=Hearings.Testimony &HearingID=127&WitnessID=482; accessed 4/2/2004)

[156]OMB Budget Documents. (http://www.whitehouse.gov/omb/budget/fy2005/index. html)

[157]OMB Budget Documents. (http://www.whitehouse.gov/omb/budget/fy2005/appendix. html)

[158]OMB Budget Documents. (http://www.whitehouse.gov/omb/budget/fy2005/appendix. html)

[159]OMB Budget Documents. (http://www.whitehouse.gov/omb/budget/fy2005/appendix. html)

[160]Homeland Security, Budget in Brief, Fiscal Year 2005.

[161]USDA VS Strategic Plan FY2004 to FY 2008, Updated February 2004.

[162]AVMA, Javma News. March 15, 2004. (http://www.avma.org/onlnews/javma/mar04/040315n.asp)

[163]USDA Fact Sheet, National Animal Health Network, May 30, 2003.

[164]USAHA Report of the Joint USAHA/AAVLD Committee on Animal Health Information Systems. 2002 Committee Reports. (http://www.usaha.org/reports/reports02/r02ahis.html)

[165]USDA Fact Sheet, National Animal Health Network, May 30, 2003.

[166]OMB's Department of Agriculture Part Assessments (Performance Measurements) February 2004. (www.whitehouse.gov/omb/budget/FY2005/pma/agriculture.pdf; accessed 4/12/04)

[167]McNair Paper 65, Chapter 4, Recommendations and Conclusions. (http//www. ndu. edu/inss/McNair/mcnair65/08_ch04.htm; accessed 4/12/04)

[168]CIDRAP. (http:www.cidrap.umn.edu/cidrap/content/bt/bioprep/news/oct0103labs. html; accessed 4/2/04)

[169]DeHaven, Ron. A Report on "Declining Infrastructure of Governmental Animal Health Professionals Puts American Agriculture at Risk." August 2003.

[170]OMB's Department of Agriculture Part Assessments (Performance Measurements) February 2004. (www.whitehouse.gov/omb/budget/FY2005/pma/agriculture.pdf; accessed 4/12/04)

[171]USDA Fact Sheet, Homeland Security Efforts, June 2003.

[172]DeHaven, Ron. A Report on "Declining Infrastructure of Governmental Animal Health Professionals Puts American Agriculture at Risk." August 2003.

[173]US Fish and Wildlife Service, Division of Law Enforcement, Annual Report FY 2001.

[174]US Fish and Wildlife, Office of Law Enforcement. www.le.fws.gov/Contact_Info_Ports.htm

[175]US Fish and Wildlife Service, Division of Law Enforcement, Annual Report FY 2001.

[176]USGS, National Wildlife Health Center. (http://www.nwhc.usgs.gov; accessed 3/8/04)

[177]USGS, National Wildlife Health Center, About the NWHC. (http://www.nwhc.usgs.gov/about_nwhc/index.html; accessed 3/8/04)

[178]USGS, NWHC, Diagnosing and Controlling Wildlife Disease. (http://www.nwhc.usgs.gov/about_nwhc/diagnosing_controlling.html; accessed 3/8/04)

[171]USGS-NWHC Information Sheet, June 2001.

[180]USGS-NWHC, Foot and Mouth Disease. (http://www.nwhc.usgs.gov/research/fmd/fmd.html; accessed 3/8/04)

[181]APHIS-VS Strategic Plan FY 2004 to FY 2008, Updated February 2004.

[182]USGS-NWHC, West Nile Virus Project. (http://www.nwhc.usgs.gov/research/west_nile/west_nile.html; accessed 3/8/04)

[183]USGS, The National Biological Information Infrastructure, an Overview, May 2001. (http://www.nbii.gov)

[184]NBII, Federal NBII Partners. (http://www.nbii.gov/about/partner/feds.html; accessed 3/18/04)

[185]NBII, Fisheries and Aquatic Resource Node. (http://www.nbii.gov)

[186]National Biological Information Infrastructure, A National Biological Information Infrastructure Overview. January 2003. (http://www.nbii.gov)

[187]NBII Fact Sheet, Wildlife Disease Information Note. November 2002. (http://www.nbii.gov)

[188]Vivian Pardo Nolan, Program Coordinator, Wildlife Disease Information Node, National Biological Information Infrastructure (NBII) Program, U.S. Geological Survey, Biological Informatics Office (Personal Communication, 4/13/04).

[189]USDA Fact Sheet, USDA Wildlife Services Protects Livestock.

[190]USDOC Seafood Inspection Program Fact Sheet. (http://www.seafood.nmfs.noaa.gov); accessed 3/17/04)

[191]US DOC. The Rationale For A New Initiative in Marine Aquaculture. September 2002.

[192]US DOC. The Rationale For A New Initiative in Marine Aquaculture. September 2002.

[193]Guide to Federal Aquaculture Programs and Services. (http://ag.ansc.purdue.edu/aquanic/jsa/federal_guide/index.htm; accessed 3/17/04)

[194]Merril, R.A., and J.K. Francer, Organizing Federal Food Safety Regulation. Seton Hall Law Review, No.1,Vol.31, 2000.

[195]CRS Issue Brief for Congress, Meat and Poultry Inspection Issues, Updated June 6, 2003. IB10082

[196]FDA Bovine Spongiform Encephalopathy, Emergency Response Plan, December 24 ,2003. (http://www.fda.gov/oc/opacom/hottopics/bseplan/bseplan.html; accessed 4/2/04)

[197]FDA Fact Sheet, The Food and Drug Administration's Seafood Regulatory Program. (http://www.cfsan.fda.gov/~lrd/sea-ovr.html; Accessed 3/17/04)

[198]Crawford, Lester, Statement before the Committee on Government Affairs, US Senate, November 19, 2003.

[199]FDA Fact Sheet, Safeguarding Animal Health to Protect Consumers. Publication No FS01-6. Revised Feb 2002.

[200]FDA Fact Sheet, CVM. (http://www.fda.gov/cvm/aboutcvm/aboutosc.htm; accessed 2/27/04)

[201]FDA Bovine Spongiform Encephalopathy, Emergency Response Plan, December 24 ,2003. (http://www.fda.gov/oc/opacom/hottopics/bseplan/bseplan.html; accessed 4/2/04).

[202]CRS Issue Brief for Congress, Meat and Poultry Inspection Issues, Updated June 6, 2003. IB10082

[203]Ibid.

[204]Merril, R.A., and J.K. Francer, Organizing Federal Food Safety Regulation. Seton Hall Law Review, No.1,Vol.31, 2000.

[205]Crawford, Lester. Statement before the Committee on Government Affairs, U.S. Senate, November 19, 2003.

[206]Merril, R.A., and J.K. Francer, Organizing Federal Food Safety Regulation. Seton Hall Law Review, No.1,Vol.31, 2000.

[207]Case, James, School of Vet Med, Univ of CA, Davis. Presentation on Agroterrorism, Public Health and HL7, Integration and Future Direction.

[208]Crawford, Lester, Statement before Committee on Government Affairs, U.S. Senate, November 19, 2003.

Appendix D

Animal Diseases and Their Vectors

The following conclusions were drawn in the NRC report *Countering Agricultural Bioterrorism* (NRC, 2003a):

1. Some animal diseases are of greater economic than public health importance. Even though significant public health impacts could in principle result from intentional introduction of animal diseases, their spread would likely be minimized by regulatory bans and procedures.
2. Limitations of current diagnostic tests and current understanding of the pathogenesis and epidemiology of specific animal diseases may make these diseases suitable for use in hoaxes.
3. Regulatory controls can substantially reduce the likelihood of natural introduction of some animal diseases.
4. Because some animal diseases are not highly contagious, selective culling would be possible if we had sensitive on-the-hoof preclinical diagnostic tools.
5. Development of effective diagnostic and identification tools for the animal diseases of concern warrants a high research priority today.
6. Basic-science and technology programs will have broad application in protecting us from harm.
7. Effective public information materials should be drafted in advance of natural or terrorist introduction of animal diseases of concern into the United States, so that they will be available immediately whenever needed.
8. The threat of an animal disease as an agricultural terrorist agent will be limited by factors such as

- Difficulty in obtaining or producing the agent.
- Physical and biological security of the plants that manufacture animal feeds, animal medicinals, and vaccines.
- Regulatory actions.
- An active national surveillance program.

9. Vulnerability to animal disease as an agricultural terrorist threat agent is increased by

- Limited effectiveness of border controls (for example, inspection procedures that are not developed with terrorists in mind, with a small proportion of luggage inspected at ports of entry).
- The small number and low sensitivity of diagnostic tests to detect an agent in living animals or animal tissues.
- A high resistance of an agent to inactivation by physical and chemical treatments.
- Lack of full compliance with regulations in place to control or eradicate the disease.
- A long incubation period from exposure to onset of disease, which would allow time for terrorists to escape detection and for wide dissemination of infected animals before discovery.
- An unwarranted degree of public concern over the disease, which could leverage a small number of cases or a hoax into an event with major adverse economic, social, and political effects.

10. Modern molecular field tests for animal diseases of concern need to be validated and introduced by USDA regionally and encouraged locally.

11. Vaccine stocks for animal diseases of concern need to be modernized and expanded.

12. Research should be performed to develop vaccines suitable for specific disease subtypes.

13. The United States should investigate the global eradication of those animal diseases posing significant threats and cooperate with international agricultural and wildlife experts in doing so. A continuing international mechanism to identify measures needed for global eradication of particular diseases should be established. Through such a mechanism, a global vaccination and eradication strategy could be developed with the participation of diverse experts and stakeholders. This could be a win-win situation for the United States and for other countries.

14. Widespread distribution of potential vector species increases the potential public health and economic impacts of a zoonotic disease.

15. It is essential for an effective response to have in place an infrastructure of disease surveillance and response systems, as well as cooperation and communication among agricultural, wildlife, and public health organizations.

16. Early detection and diagnostic tools are pivotal for limiting the extent of an outbreak. Education, limitation of animal movement, and development of vector population control methods are other important factors.

17. Basic research is critical for understanding of the pathogenesis and epidemiology of many animal diseases.

Appendix E

Biographical Sketches of Committee Members

Lonnie J. King, *Chair,* is the Dean of the College of Veterinary Medicine at Michigan State University and the Director of the Office of Strategy and Innovation at the Centers for Disease Control and Prevention (CDC). Dr. King previously held various positions in the government, as Administrator of the USDA Animal and Plant Health Inspection Service (APHIS) and Associate Administrator and Deputy Administrator for USDA/APHIS/Veterinary Services. Before his government career, Dr. King was a private practitioner and had experience as a field veterinary medical officer, station epidemiologist, and staff assignments involving Emergency Programs and Animal Health Information. Dr. King directed the American Veterinary Medical Association's Office of Governmental Relations and is certified in the American College of Veterinary Preventive Medicine. He is the past President of the American Association of Veterinary Medical Colleges, current President of the American Veterinary Epidemiology Society, and has served as Co-Chair for both the National Alliance for Food Safety and the National Commission on Veterinary Economic Issues. He is the Lead Dean at Michigan State University for food safety with responsibility for the National Food Safety and Toxicology Center. He is also codeveloper and course leader for Science, Politics, and Animal Health Policy. Dr. King received his B.S. and D.V.M. degrees from Ohio State University, and his M.S. degree in epidemiology from the University of Minnesota. He also attended the Senior Executive Program at Howard University and received a M.P.A. from American University. Dr. King was elected to the Institute of Medicine in 2004. He has served on the NRC Committee on Ensuring Safe Food from Production to Consumption, the Committee on Opportunities in Agriculture, and the Steering Committee for a Workshop on the Control and Prevention of Animal Diseases.

Margaret Hamburg, *Co-Chair*, is Vice President for Biological Programs at the Nuclear Threat Initiative (NTI). Dr. Hamburg is a physician and expert in public health and bioterrorism. Before joining NTI, Dr. Hamburg was assistant secretary for planning and evaluation at the U.S. Department of Health and Human Services. Dr. Hamburg was the commissioner of health for the City of New York and former assistant director of the Institute of Allergy and Infectious Diseases at the National Institutes of Health. She is a member of the Institute of Medicine, the New York Academy of Medicine, the Council on Foreign Relations, and a fellow of the American Association of the Advancement of Science. Dr. Hamburg is currently serving on several committees at the National Academies, including the Roundtable on Scientific Communication and National Security, the Committee on International Security and Arms Control, and the Working Group on Biological Weapons Control (Chair), and as a member of the Board on Global Health. She previously served on the committee on Science and Technology for Countering Terrorism and the Committee on Emerging Microbial Threats to Health in the 21st Century.

Sharon Anderson is currently Director of the North Dakota State University Extension Service located in Fargo. Dr. Anderson assumed that role in January 1995. Her previous experience with the NDSU Extension Service, which began in 1970, includes serving as a district director, program leader for youth and family, 4-H youth development specialist, and family and consumer science specialist. Dr. Anderson received her Ph.D. from the University of North Dakota in higher education administration. Dr. Anderson served on the Extension Committee on Organization and Policy from 1996 to 1999 and was chair in 1999. She has been on the National 4-H Council Board of Trustees since 1997 and completed a term as vice chair in 2002.

Corrie Brown is a Professor in the College of Veterinary Medicine, University of Georgia. Her research interests include pathogenesis of infectious disease in food-producing animals through the use of immunohistochemistry and in situ hybridization. She is active in the fields of emerging diseases and international veterinary medicine and currently serves as Coordinator of Activities for the College of Veterinary Medicine. Dr. Brown completed a D.V.M. (University of Guelph), followed by a Ph.D. (University of California at Davis) in veterinary pathology, specializing in infectious diseases. Prior to joining University of Georgia in 1996, she worked at the USDA Plum Island Foreign Animal Disease Center for 10 years, conducting pathogenesis and control studies on many of the foreign animal diseases. Her bench research interests at University of Georgia have been focused on poultry diseases, and she works closely with the USDA facility in Athens that is dedicated to foreign diseases of poultry. In educational research, she has several grants to help promote awareness of foreign animal diseases and global issues in veterinary curricula and beyond. Dr. Brown is currently serves on the Committee on Genomics Databases for Bioterrorism Threat Agents: Striking a Balance for Information Sharing.

Tim Herrman is a Professor at Texas A&M University where he serves as State Chemist and Director, Office of the Texas State Chemist. Prior to assuming this responsibility in December 2004, Dr. Hermman was a Professor in the Department of Grain Science and Industry at Kansas State University, where he served as the Extension State Leader and Director of Graduate Studies. He chaired the American Feed Industry Association's Quality Council in 2004 and is on the executive committee of two national research projects that address food safety, security, and marketing. Dr. Herrman has published numerous articles and extension bulletins and runs a nationally recognized research program. Before pursuing his Ph.D., he worked 5 years with Anheuser-Busch Inc. as the coordinator of field operations, purchasing potatoes and barley in the western United States. Dr. Herrman received his bachelor's degree in agronomy at Washington State University and his master's degree in plant pathology and doctorate in plant science at the University of Idaho.

Sharon K. Hietala is a professor of clinical diagnostic immunology with the California Animal Health and Food Safety Laboratory System, and has a joint appointment in the School of Veterinary Medicine, Department of Medicine and Epidemiology at the University of California, Davis. Dr. Hietala earned a bachelor's degree in bacteriology in 1976, and a Ph.D. in comparative pathology in 1987, both from UC Davis. Sharon joined the California Animal Health and Food Safety Laboratory in 1989, where she is responsible for the immunology and biotechnology services in the five-laboratory system. Her professional interests include serology, molecular diagnostics, and diagnostic epidemiology. She serves on the USDA National Surveillance System Steering Committee and is active in the American Association of Veterinary Laboratory Diagnosticians, the U.S. Animal Health Association, and a variety of food animal and poultry industry issue and interest groups.

Helen H. Jensen is Professor of Economics and Division Head of Food and Nutrition Policy Research in the Center for Agricultural and Rural

Development at Iowa State University. Her current research focuses on food and nutrition programs and policies, issues related to food security and safety, including the economics of food safety, food systems and hazard control options, and animal diseases. Dr. Jensen currently serves on the editorial board of the *Journal of Agricultural Economics* and has been an active member of the American Agricultural Economics Association, where she has chaired several working committees. She was a member of the National Research Council's Panel on Animal Health and Veterinary Medicine from 1995 to 1998, and a member of the Committee on Biological Threats to Agricultural Plants and Animals. She currently serves on the Committee on National Statistics' Panel to Review USDA's Measurement of Food Insecurity and Hunger, and she recently served on the Institute of Medicine's Committee to Review the WIC Food Packages. She joined the faculty at Iowa State in 1985 and holds a Ph.D. degree in Agricultural Economics from the University of Wisconsin.

Carol A. Keiser is president of C-BAR Cattle Company, Inc., where she established and manages operations for the feeding of 5,000 head of cattle in feedlots in Texas, Kansas, Nebraska, and Western Illinois. Prior to C-BAR Cattle Company, she developed health and well-being procedures for Loveless Feedlot. She is a Certified Livestock Manager, Certified Livestock Dealer and was elected to the Board of Directors of the Council of Food and Agricultural Research (C-FAR). She is the past Chair of UIAA University of Illinois Alumni Association and Board of UIUC Agriculture, Consumer and Environmental Sciences Alumni Board. She is a graduate of the University of Illinois in Animal Science and a professional member of the American Society of Animal Science and American Meat Science Association. Currently, Ms. Keiser represents food animal commodity producers on the National Agricultural Research, Extension, Education, and Economics Advisory Board. Additionally, she serves as Chair of the Farm Foundation, Roundtable Steering Committee.

Scott R. Lillibridge is Professor of Epidemiology and Director, Center for Biosecurity and Public Health Preparedness at The University of Texas Health Science Center at Houston. Most recently, he worked as Special Assistant for National Security and Emergency Management for the Secretary of the Department of Health and Human Services and assisted in the development of a national bioterrorism program at HHS during a time when the nation was experiencing anthrax attacks in October 2001. Previously, he developed and was the founding Director of the Bioterrorism Preparedness and Response Program at the Centers for Disease Control and Prevention (CDC) starting in 1998. This office was charged with enhancing state and local capacities to respond to bioterrorism. In addition to infectious disease concerns, other CDC efforts in support of this program included consideration for chemical terrorism, a national pharmaceutical stockpile, health communication, training and national lab enhancement. His career at CDC focused on emergency public health response issues. He was the lead physician during the initial United States Public Health Service (PHS) response to the Oklahoma City bombing and also led the U.S. Medical Delegation to Tokyo following the sarin release in 1995. During the 1996 Olympics, he served as the HHS Science Advisor to the multiagency task force that was assembled to protect the public against biological and chemical terrorism. He has worked in emergency response and preparedness roles throughout the world in support of the United States government and nongovernmental organizations. Dr. Lillibridge was recently appointed by President George W. Bush to the White House Emergency Services, Law Enforcement, and Public Health and Hospitals Senior Advisory Committee for Homeland Security. Dr. Lillibridge earned his B.S. form East Tennessee State University and his M.D. from the Uniformed Services University of the Health Sciences, F. Edward Hebert School of Medicine.

Terry McElwain is the Executive Director of the Washington Animal Disease Diagnostic Laboratory and Director of the Animal Health Research Center in the College of Veterinary Medicine at Washington State University. He is Past President of the American Association of Veterinary Laboratory Diagnosticians and has been a key architect in the creation and development of the National Animal Health Laboratory Network. He interacts frequently with the Centers for Disease Control and is also a member of the governor's emergency preparedness task force in the state of Washington. Dr. McElwain has a long and established research record in the field of veterinary infectious diseases, especially those of agricultural animals. He received his D.V.M. from the College of Veterinary Medicine, Kansas State University, in 1980, and his Ph.D. from Washington State University in 1986.

N. Ole Nielsen, a professor emeritus and former dean of the Ontario Veterinary College (1985-94) and the Western College of Veterinary Medicine, University of Saskatchewan, has particular interests in comparative medicine and ecosystem health. He attended the University of Toronto, receiving his D.V.M. in 1956, and the University of Minnesota, where he received his Ph.D. in veterinary pathology in 1963. He joined the University of Saskatchewan in 1964 and subsequently became Dean of the Western College of Veterinary Medicine (1974-1982). He has served on the Boards of a number of research agencies including; the Veterinary Infectious Disease Organization (VIDO), University of Saskatchewan; Canadian Centre for Toxicology; Association of Canadian Universities for Northern Studies (ACUNS); International Laboratory for Research on Animal Disease (ILRAD), Nairobi; and the International Livestock Research Institute (ILRI), Nairobi. His interests in environmental issues and the development of the concept of ecosystem health are reflected in: service as chair of the Saskatchewan Environmental Advisory Council, (1978-1982); symposia planning for the International Society for Ecosystem Health (ISEH) in Ottawa (1994), Copenhagen (1996), Sacramento (1999); promoting research in agroecosystem health at the University of Guelph; and involvement in the Ecosystem Approaches to Human Health program of the International Development Research Centre. He was president of the Canadian Veterinary Medical Association in 1969. Dr. Nielsen currently serves on the National Research Council's Board on Agriculture and Natural Resources.

Robert A. Norton is a Professor at Auburn University. A microbiologist by training, he was educated at Southern Illinois University, where he received his B.S. and M.S. Dr. Norton served in the U.S. Army Chemical Corps and later with the United States Army Medical Research Institute of Infectious Diseases at Fort Detrick, Maryland, where he worked on projects including the development of novel vaccines for botulism and other bacterial pathogens. After his service, Dr. Norton moved to the University of Arkansas, where he earned a Ph.D. in Veterinary Bacteriology. He has been a member of the faculty at Auburn since 1995. He presently serves as research leader in the Poultry Microbiology and Parasitology Laboratory and conducts research on bacterial diseases in animals. Dr. Norton also works on the issues of agricultural bioterrorism defense and currently serves as the editor for *Issues in Ag-Security*, a monthly e-mail newsletter that is sent to subscribers in government and industry.

Michael T. Osterholm is the Director of the Center for Infectious Disease Research and Policy (CIDRAP) at the University of Minnesota, where he is also Professor, School of Public Health. He is also serves as the Associate Director of the Department of Homeland Security's National Center for Food Protection and Defense and was recently appointed to the newly created National Science Advisory Board on Biosecurity. Previously, Dr. Osterholm was the state epidemiologist and Chief of the Acute Disease Epidemiology Section for the Minnesota Department of Health. Following the September 11 terrorist attacks, Dr. Osterholm has served as a special advisor to the U.S. Secretary of the Department of Health and Human Services on issues related to bioterrorism and public health preparedness. He has received numerous awards from the National Institute of Allergy and Infectious Diseases and the Centers for Disease Control and Prevention (CDC). He served as principal investigator for the CDC-sponsored Emerging Infections Program in Minnesota. He has published more than 300 articles and abstracts on various emerging infectious disease problems and is the author of best selling book *Living Terrors: What America Needs to Know to Survive the Coming Bioterrorist Catastrophe.* He is past president of the Council of State and Territorial Epidemiologists. He is a member of the Institute of Medicine and currently serves on the Institute's Forum on Emerging Infections. He also served on the Committee to Ensure Safe Food from Production to Consumption.

Patricia Quinlisk is a medical epidemiologist practicing at the Iowa Department of Public Health, where she functions as both the Medical Director and the State Epidemiologist. Her background includes training as a clinical microbiologist (MT[ASCP]), training microbiologists while a Peace Corps Volunteer in Nepal, a Masters of Public Health from Johns Hopkins (with a emphasis in infectious disease epidemiology), a medical degree from the University of Wisconsin, and training as a field epidemiologist in the Centers for Disease Control and Prevention's (CDC's) Epidemic Intelligence Service. Every year, for the last 12 years, she has conducted weeklong epidemiologic training courses in Europe, and as a professor at the University of Iowa and Iowa State University, lectures regularly at other educational institutions around the Midwest. She serves, or has served, on several national advisory committees including the National Vaccine Advisory Committee, the Sub-Committee for Vaccine Safety and Communication, the Advisory Committee of the U.S. Marine Corps Chemical/Biological Incident Response Force, the Department of Defense's Panel to Assess the Capabilities for Domestic Response to Terrorist Acts Involving Weapons of Mass Destruction (the Gilmore Commission), and as President of the Council of State and Territorial Epidemiologists (CSTE). She has testified before two Congressional Subcommittees on Public Health aspects of terrorism, and participated on the Institute of Medicine's Committee on Microbial Threats to Health in the 21st Century and its Committee on the Psychological Consequences of Terrorism. Recently, she was named to the Board of Scientific Counselors for the National Center for Infectious Diseases, Centers for Disease Control and Prevention.

Linda Saif is a professor and researcher with Ohio State University's Ohio Agricultural Research and Development Center (OARDC), working on the mechanisms in immunity against intestinal infections. Dr. Saif's research focuses on enteric viruses, including rotaviruses, caliciviruses, and coronaviruses, which cause mortality and morbidity in both food-producing animals and humans. During the past 30 years, she has identified new intestinal viruses and developed diagnostic tests and research methods for working with them in the laboratory. Furthermore, she discovered viruses that cause intestinal diseases in livestock and developed methods for their control. Her contributions to mucosal immunology and intestinal virology have had major impacts on animal and human health research and vaccine development. She is also credited with discovering the potential of enteric viral infections in animals to infect human populations in epidemic proportions. One example is Dr. Saif's ongoing effort to develop safe and effective vaccines for rotavirus diarrhea, which kills nearly one million children every year. Dr. Saif earned her bachelor's degree from the College of Wooster in 1969 and received her master's degree (1971) and doctorate (1976) in microbiology/immunology from Ohio State. She has been an OARDC faculty member since 1979, garnering more than $14 million in research grants and publishing numerous articles in books and professional journals. In 2002, Dr. Saif became the first Ohio State researcher not based on the Columbus campus to be recognized as a Distinguished University Professor, and was awarded an honorary doctorate by Belgium's Ghent University. She is an elected member of the National Academy of Sciences.

Mark Thurmond is a professor of medicine and epidemiology at the School of Veterinary Medicine at UC Davis. Dr. Thurmond has 33 years of experience as a clinician and clinical epidemiologist, mainly involving dairy cattle. Dr. Thurmond's interests relate to the epidemiology of infectious diseases of cattle and to application of epidemiological principles to prevention, control, and eradication of diseases and infections that affect animal health and productivity. Diseases of special interest include bovine viral diarrhea, neosporosis, diseases of the mammary gland, abortion, and foreign animal diseases, such as foot-and-mouth disease. Research interests in epidemiologic methods relate to diagnostic epidemiology, particularly population-based diagnostic approaches, modeling, surveillance, and diagnostic screening. Dr. Thurmond received his D.V.M. from the University of California, Davis in 1972, and the M.P.V.M. from UC Davis in 1975. He received a Ph.D. from the University of Florida in 1982.

Kevin D. Walker is Division Director of Agricultural Health and Food Safety at the Inter-American Institute for Cooperation in Agriculture. In his role as director of one of the four technical division, he has established programs in food safety leadership, financial and technical support structures for emerging health issues, networks for developing countries to gain information on trade, emerging diseases and pathogens, tracking of international trade opportunities and constraints due to sanitary and phytosanitary health standards, and integration of country-specific food safety needs with financial institutions and support agencies. Dr. Walker has also presented several analyses within the World Trade Organization on the application of sanitary and phytosanitary standards and has worked extensively with the World Organization of Animal Health (OIE) in the development of a performance, vision, and strategy instrument to enhance the modernization of national veterinary services across the world. He is formerly a director of the Center for Emerging Issues at the Animal Plant Health Inspection Service and a strategic and economic analyst at Farmland Industries, at one time the largest farmer-owned cooperative within the United States. Dr. Walker received his Ph.D. from the University of Missouri in 1985.